PRACTICAL ASSESSMENT

PHYSICS

for GCSE

John Skevington

UNWIN HYMAN

Published by
UNWIN HYMAN LIMITED
15–17 Broadwick Street
London W1V 1FP

British Library Cataloguing in Publication Data

Skevington, John
    Practical Assessment: Physics for G.C.S.E.
    1. Physics – Study and teaching
    (Secondary) – Great Britain.
    2. Curriculum-based assessment.
    3. General Certificate of Secondary Education.
    I. Title
    530'.076   QC47.G7

ISBN 0 7135 2823 0

Illustrated by RDL Artset Ltd
Cover design by Charles Snape

Typeset by Nene Phototypesetters Ltd, Northampton
Printed in Great Britain by
Bell & Bain Ltd., Glasgow

530

93-24

Contents

Acknowledgements

Few books are solely the work of the author and this one is no exception. I am grateful to David Edwards who originally suggested that I write a book on the assessment of practical work and to Professor David Waddington, whose assistance when it seemed that the original project might come to naught was most timely. My ideas about assessment have developed as I have discussed the problems with colleagues and teachers from other schools, and if they recognise some of their ideas in this work then I hope that they will accept my thanks.

Both Christopher Blake and Pat Winter at Unwin Hyman have been of tremendous help throughout the production of this book and their helpful comments and careful attention to detail have eased the process considerably.

The person who has borne most of the burden during the writing of this book has been my wife Anne, whose encouragement and support have proved invaluable. Anne and our children have endured much during the eighteen months that it has taken to finish it and I hope that the result goes some way towards compensating them.

John H. Skevington
Leeds 1987

1 Why internal assessment?

The place of practical work in school physics courses

Few teachers would quarrel with the view that "practical work" holds a central position in the teaching of school physics. The reasons given by teachers for this prominence include the promotion of student motivation and interest, the illustration and support of theory, the development of the skills and techniques which are associated with practical work, and the idea that practical work allows pupils to "behave like a scientist".

In spite of these aims, the theoretical content of physics has influenced courses to such an extent that the role of practical work as a support of theory has tended to dominate. Given this, it is hardly surprising that the major emphasis in assessment has been directed towards the knowledge and recall of content rather than on the ability to carry out experiments which support it.

The 1970s saw the beginnings of a change in the way in which school science (and hence physics) is viewed. The Schools Council Science Project (SCISP) placed an increased emphasis on the processes of science and a concomitant reduction in the importance of content. This move is echoed in the APU model which sees science as "an experimental subject concerned fundamentally with problem solving". As a consequence, the APU monitoring laid emphasis "on particular processes and skills which should be the outcomes of science education".

Whether or not one agrees with the APU model, there is no denying that it has had a significant effect on the National Criteria for GCSE in the Sciences. These have accepted the view that science courses have become overburdened by content as they have tried to keep pace (albeit fifty or sixty years behind) with the explosive growth of science and technology. The result is a declared intention to reduce the content of physics syllabuses so that more time can be spent on the development of skills and processes.

The place of practical skills in the National Criteria

Both the National Criteria for Science and for Physics contain statements of the aims which stress the importance of practical work and the development of related skills in GCSE courses. These aims are listed in Table 1.1.

The outcome of these aims (and the related assessment objectives which will be considered in the next chapter) is that the examining groups no longer have the freedom to choose whether or not the assessment of practical work is to form part of the examination scheme.

The place of practical assessment in the National Criteria

There are a number of important innovations in the National Criteria which are having a significant impact on physics teaching. Most important from the point of view of coursework assessment is the principle of "fitness for purpose" which requires that all examination components and assessment procedures should reflect and be appropriate to the nature of the subject, its educational aims and assessment objectives.

The principle recognises that written tests of material related to practical and experimental skills are not an adequate measure of attainment. This echoes the findings of the APU which suggest that "pencil and paper analogues of 'practical' questions can produce results which differ from those of the practical questions themselves" (APU, 1985).

As a result of this, all schemes must allocate not less than 20% of the total marks to experimental skills and at least 10% on the basis of experimental and observational work in the laboratory.

Given that the assessment of practical work is to be an integral part of physics courses, teachers need to

Table 1.1 *Aims of courses in Physics which are relevant to practical work*

From the National Criteria for Science	From the National Criteria for Physics
2.1 To provide through well designed studies of experimental and practical science a worthwhile educational experience for all pupils . . . and in particular, to enable them to acquire sufficient understanding and knowledge.	2.6 To provide an appropriate body of knowledge, understanding and skills . . .
2.2 To develop abilities and skills that: 2.2.1 are relevant to the study and practice of science; 2.2.3 encourage safe practice.	2.7 To develop the skills of observation, experimentation, processing and interpretation of data, evaluation of evidence and formulation of generalisations and models.
2.3 To stimulate: 2.3.1 curiosity, interest and enjoyment in science and its methods of enquiry.	2.8 To encourage students to apply, qualitatively and quantitatively, their knowledge and understanding of physical principles to familiar and unfamiliar situations.
	2.9 To ensure that students: 2.9.1 can follow instructions 2.9.2 understand the need for, and comply with, safety precautions
	2.10 To foster relevant communication skills

adopt a positive attitude towards these changes. Teachers who already have experience of operating schemes of internal assessment of practical work have found that they can:

– increase understanding of practical work;
– provide feedback of each student's progress;
– allow assessment of performance of the processes of practical physics and not just the outcomes;
– ease the pressures and anxieties felt by students in terminal examinations; and
– increase the motivation for practical work and thus enhance the learning process.

Furthermore, teachers who have been involved in Mode 3 schemes of assessment would agree that coursework assessment enhances the status of teachers, by recognising their professional competence and responsibility.

The important thing to remember is that our main concern should be the teaching and development of practical skills in our pupils. Assessment is a means of measuring pupil attainment. Coursework assessment is a means of determining pupil progress. Seen in these terms, the introduction of GCSE and the changes it brings can be important tools for increasing the effectiveness of our teaching in what, all agree, is an important area of physics.

2 What needs to be assessed?

Introduction

In the first chapter, we considered the educational aims which are relevant to practical work. However, it is the assessment objectives which define the shape of the practical work which will form the subject of the assessment.

The examining groups have developed schemes of assessment from a consideration of the assessment objectives specified by the National Criteria. The syllabi have all grouped the processes involved in experimental work into a number of skill areas. They have provided descriptions of performance (with varying degrees of detail) which are intended to enable teachers to develop exercises and "mark schemes" for the purposes of assessment.

In this chapter we will consider these skill areas and the performance descriptions in an attempt to clarify the method of distinguishing between different levels of performance. In particular, we will consider those aspects of the performance descriptions which do discriminate between levels so that teachers will be able to focus more finely on the elements which an exercise should involve if it is to provide candidates with the opportunity to display ability at all levels of performance.

The assessment objectives

There are a number of objectives in both the Science and the Physics Criteria which could be considered as being directly relevant to the science skills and processes which are involved in experimental work. These are shown in Table 2.1. Since physics syllabi must fulfil the Science Criteria, the schemes of practical assessment have been based on both sets of objectives.

Given that all schemes must conform to these criteria, it is somewhat surprising that there is such a wide variation in those which have been produced. The variation is such that it is not possible to provide a single set of analyses of skill areas which is applicable to every syllabus. As we have seen, the National Criteria contain a large number of relevant assessment objectives. There is some overlap between these objectives and each examining group has interpreted them in slightly differing ways so that, although skill areas in different syllabi may bear the same title, they do not necessarily cover the same objectives.

The schemes put forward by the four English boards do however have some similarities and these are summarised in Table 2.2.

Most syllabi have provided rather general descriptions of the standards of performance to be expected for each skill area, although the NEA and the SEG have adopted definite criteria for the award of marks. Even with criteria however, a measure of interpretation and clarification is required. Such an analysis helps in understanding the mark scheme, which assists in maintaining uniform standards, and is essential for the design of effective exercises.

An attempt to provide a comprehensive survey would involve either a separate set of analyses for each syllabus, or an analysis of each objective (or group of

Table 2.1 *Objectives from the National Criteria which relate to the assessment of practical and experimental skills*

SCIENCE	PHYSICS
candidates are expected to demonstrate the skills and abilities to	candidates are expected to
3.2.1 observe, measure and record accurately and systematically	
3.2.2 follow instructions accurately for the safe conduct of experiments	3.2.5 show understanding of safety procedures
3.2.3 communicate scientific observations, ideas and arguments logically, concisely and in various forms	3.3.7 present information in a precise and logical form
3.2.4 translate information from one form to another	3.3.5 translate information from one form to another
3.2.5 extract from available information data relevant to a particular context	
3.2.6 use experimental data, recognise patterns in such data, form hypotheses and deduce relationships	
3.2.7 draw conclusions from and evaluate critically, experimental observations and other data	3.3.9 draw conclusions and formulate generalisations
3.2.8 recognise and explain variability and unreliability in experimental measurements	3.3.8 recognise mistakes, misconceptions, unreliable data and assumptions
3.2.9 devise and carry out experimental and other tests to check the validity of data, conclusions and generalisations	3.3.4 solve problems by designing, conducting and interpreting the results of simple experiments
3.2.10 devise and carry out experiments or other tests for particular purposes, selecting suitable apparatus and using it effectively and safely	
3.2.13 apply scientific ideas and methods to solve qualitative and quantitative problems	
3.2.14 make decisions based on the examination of evidence and arguments	
3.2.15 recognise that the study and practice of science are subject to various limitations and uncertainties	

Table 2.2   An analysis of the schemes adopted for assessment of coursework by the four English examination groups

	LEAG A	LEAG B	MEG incl. Nuffield	NEA A & B	SEG incl. Alternative syllabus
Skill areas	Manipulative skills	Manipulative skills	Use and organisation of apparatus	Following instructions Selecting apparatus	Setting up the experiment
	Observational skills	Observational skills Measurement skills	Measurement and observation	Making quantitative observations	Making observations
	Recording and Reporting	Use and analysis of data Recording and Reporting	Handling experimental data	Presentation of results Drawing conclusions from experimental evidence	Recording and processing
	Design, planning and execution	Designing and planning	Design and execution of experiments	Design and carry out an experiment to solve a problem	Design the experiment
Mark range	0–5	0–5	1–10	0–3	various
Number of assessments	At least 2 per area	At least 2 per area	At least 2 per area	At least 2 per area	On basis of 4 experiments
Weighting	20%	20%	20% Nuffield 28·4%	20%	30%
Moderation	Inspection by assessor	Inspection by assessor	External on sample of scripts	Statistical plus visiting "coordinators"	By training teachers to standard

closely related objectives) which the teacher can use to assemble appropriate performance descriptions for a particular syllabus. This latter approach would seem to be the most effective and has been adopted here. The objectives have been grouped under general headings but these do not necessarily correspond to those of any particular syllabus.

As a starting point we will analyse the performance descriptions provided by each syllabus in terms of the "discriminators". Discriminators are the common factors which are used to separate the different levels of performance. A good understanding of these discriminators is the first stage in designing exercises which will enable us to make valid and acceptable assessments.

The "amount of help" – a general point

A number of syllabi specify the "amount of help" required by the pupil as a discriminator in a number of the skill areas. There are a number of problems which arise from this and we will consider them here.

It would be inappropriate to attempt to make assessments of the relevant skill areas when candidates are working in groups (for those syllabi which allow group work). Even when candidates are working individually, but carrying out the same exercise or exercises, they will always be able to gain a measure of help from observing the performance of other candidates.

The problem has always existed in practical examinations and, unless we take the extreme and unacceptable step of testing individual candidates in isolation, there is little we can do to eliminate it. However, the problem is more acute in the context of GCSE assessment where the amount of help is a criterion of performance. It is obviously unfair to penalise a candidate who has been given help by the

teacher when those around will "get it for free"!

There are two aspects to the problem. Candidates who observe performance and act on their observations may be obvious and the teacher must make a judgement as to whether or not the assessment of the skill area involved should continue. On the other hand, candidates may overhear the teacher talking and it would be expecting too much of human nature to assume that they would not take advantage of any information which they might receive. In order to minimise the chances of this happening, the use of help sheets might be considered. These are especially appropriate in the area of experimental design, and examples are provided in the sample exercises for this skill area.

The other important problem raised by the use of help as a discriminator is the danger that we may tend to measure it against the norm for the group. We must have a more absolute scale if there is to be comparability of standards between different teachers and centres.

I would suggest a coarse three-point scale to discriminate between the different levels.

Low (significant or considerable help) — The candidate has to be told or shown what to do at most stages in the experiment

Medium (little guidance) — Intervention may be necessary at some stage in the exercise to avoid mishap or ensure continuation. Help may be more in the nature of prompting rather than instruction.

High (no help or guidance) — No intervention on the part of the teacher, although probing questioning might be employed to draw out the pupil's ideas.

It is a matter of professional judgement as to the nature and quantity of help given, but the above guidelines should enable the teacher to make more objective assessments.

The problems raised by the use of help as a discriminator are such that it is important that the examining groups consider it as a matter of urgency when they review their syllabi.

Following instructions and manipulation
(Objectives 3.2.2 and 3.2.10 in Science)

All of the various syllabi include references to the ability to follow instructions in some aspect of their performance criteria, although only the NEA syllabi assign one skill area exclusively to this skill. In the other schemes, the ability to follow instructions forms a part of the skill area of manipulation or use and organisation of apparatus.

It would seem sensible to combine the assessment of the ability to follow instructions with manipulation since the observation of a candidate's performance is likely to involve aspects of both skills. In some schemes, reference is made only to the ability to follow instructions in the lower levels of performance criteria, as in the LEAG Syllabus A. The ability is implicit however, in all schemes, as, without it, candidates are unable to carry out any exercise.

Table 2.3 lists the discriminators adopted by each scheme.

Table 2.3 Discriminators used in the area of following instructions, manipulation and selection of apparatus

LEAG "A"	
Manipulative Skills	Amount of help in handling/selecting apparatus
LEAG "B"	
Manipulative skills	Length of sequence of instructions and nature of task
MEG	
Use and organisation of apparatus	Amount of help, length of sequence of instructions, dexterity and confidence in handling apparatus
NEA	
Following instructions	Amount of help
Selecting apparatus	Suitability in terms of accuracy, sensitivity and safety
SEG	
Setting up equipment	Amount of help

The instructions given may be oral, written or diagrammatic. The MEG scheme suggests that the instructions may refer to the use of unfamiliar apparatus or to the carrying out of a sequence of activities with familiar apparatus, and the LEAG Syllabus B also refers to the ability to follow a complex sequence of instructions as being a suitable measure of a candidate's performance for 4 marks.

In designing suitable exercises to test this skill, the MEG syllabus provides a suitable starting point. It is difficult to envisage an adequate range of unfamiliar apparatus with which a group of candidates might be tested, and the construction of a set of exercises which provide candidates with sequences of instructions seems to be a more profitable approach.

It is possible to provide a range of exercises with increasingly complex sequences of operations, so that candidates of all ability levels are provided with the opportunity to display what they can do. At the simplest level this might involve setting up the apparatus from a pictorial diagram, at the intermediate level from a representational diagram, such as a conventional section diagram, and at the highest level the diagram could be a symbolic representation, such as a circuit diagram.

The complexity of the assembly can also be graded, either by the nature of the apparatus involved or by the complexity of the arrangement. For example, the setting up of a ripple tank and its associated equipment is a more complex procedure than the assembly of the apparatus required to measure the boiling point of water. Similarly, a series circuit is simpler to set up than a parallel circuit.

None of the syllabi specify the range of manipulative abilities which it is expected that candidates will be able to carry out (unlike the NEA Chemistry Syllabus for instance). The SEG scheme does include a list of the types of practical work of which candidates should have experience in its guidance for teachers, but this is very general. Table 2.4 gives a list of suggested abilities which were used in the design of the exercises in the resource sheets and which should prove useful for teachers who wish to develop further exercises.

Table 2.4 A list of the techniques and skills which one would expect candidates following a GCSE course in Physics to acquire

Candidates should be able confidently, correctly and safely to:

- use safety glasses;
- use a Bunsen burner;
- use a clamp stand;
- transfer liquids from one container to another;
- handle hot objects and containers of liquid;
- connect and operate a power supply;
- set up a ray box to produce a narrow beam of light;
- use pins to "shoot" rays;
- use the method of no parallax to find the position of an image;
- set up a convex lens or concave mirror to produce a focussed image;
- connect up an electrical circuit observing the correct polarity of components, meters and power supply;
- connect a potential divider in a circuit;
- set up a CRO to show a trace;
- use a plotting compass to investigate the shape and direction of a magnetic field.

In addition there are a number of skills associated with the correct use of measuring instruments which are not directly related to the ability to read the scale. These include:

- avoidance of parallax error;
- reading to the bottom of the meniscus;
- setting up a lever arm balance;
- correct positioning of a thermometer;
- starting a stopwatch at the correct time.

Some of the skills are to be expected from candidates at all levels but a few are applicable only to the higher abilities. Using optical pins to shoot rays or the method of no parallax are quite difficult techniques, and it is unlikely that Grade F candidates would be able to perform such tasks satisfactorily. Similarly, it is suggested in some syllabi that a treatment of the potential divider is suitable only for candidates aspiring to the higher grades. In such a case it would not be desirable to use exercises involving this device to assess lower ability pupils. The teacher will need to refer to the adopted syllabus to check points such as these.

Having decided on the skills and techniques which are to be used in the construction of exercises, the teacher needs to be certain of what constitutes a good performance for the purposes of assessment. We are all used to making subjective assessments of a pupil's performance during practical work – we do it every time we decide to step in to avoid disaster! If we are, however, to make assessments to a consistent standard, as well as maintaining uniformity with colleagues who are assessing other groups, we need to define successful performance more closely.

Eglen and Kempa (1974) proposed a breakdown analysis which outlined the aspects of performance which should be considered when assessing manipulative skills in chemistry. It provides us with a general approach to the assessment of manipulative skills and is equally applicable to physics.

Table 2.5   A breakdown analysis of manipulative skills (Eglen & Kempa, 1975)

Skill competence	Performance features
Methodical working	Correct sequencing of tasks forming part of an overall operation. Effective and purposeful utilisation of equipment. Efficient use of working time. Ability to develop an acceptable working procedure on the basis of limited instruction.
Experimental techniques	Correct handling of apparatus and materials. Safe execution of an experimental procedure. Taking of adequate precautions to ensure reliable observations and results.
Manual dexterity	Swift and confident manner of execution of practical task. Successful completion of an operation or its constituent part tasks.
Orderliness	Good utilisation of available working (laboratory bench) space. Organisation in the placing of equipment and materials used. Tidiness of the working area.

The performance features fall into two groups, those which are based on the ability to follow instructions – what might be termed the externally imposed parameters – and those internal aspects which the candidate contributes to the performance.

In the first group are the successful completion of an operation or its constituent parts and the correct sequencing of tasks. The remaining features are concerned with the manner of execution – such things as orderliness and purposeful utilisation of equipment.

There are also pointers to a higher level of performance. The development of an acceptable working procedure and taking precautions to ensure reliable observations are aspects which feature in the higher skill levels of both the LEAG and MEG schemes.

For an objective assessment of each of the "performance features" Kempa (1986) advocates the elaboration of each feature in the context of the practical task which is to be assessed. Examples based on items from Table 2.4 are given below.

Clamp	– all screws adjusted so that there is no play in any part of the assembly
	– boss and clamp positioned so that assembly is stable
	– clamp holding supported object firmly and safely
	– safe position on bench
Power supply	– plugs placed in correct terminals
	– wires securely and safely fixed to terminals if used
	– output checked or reduced to a low level before switching on
	– safe positioning of apparatus, including position of flex
	– switched on after checking above points
CRO	– switched on and allowed to warm-up with brightness set at low level
	– beam focused and brightness set at a reasonable level
	– spot centred
	– Y-gain set at sensible level
	– time-base set at sensible value to give meaningful trace (if the time-base is to be used)
	– safe positioning of instrument

The number of operations involved in the performance of even the most basic of practical tasks is quite large when it is analysed in this way. If the teacher were to attempt to use such a check-list as a mark scheme for the assessment of every candidate, the process would take an unacceptable length of time. The analysis does, however, focus attention on the skills to be developed and it can help to ensure that different teachers are assessing to a comparable standard.

The nature of the skill means that assessment will involve a significant amount of direct observation. In addition, pupils will need to be given the opportunity to work individually during the part of the exercise where the assessment is to take place. This puts restrictions on the types of strategy which will be available to us.

Observation and measurement
(Objectives 3.2.1 and 3.2.5 in Science)

Objective 3.2.1 covers three quite distinct skills and, although some schemes have included all three within one skill area, others have separated them into three skills or considered two of them. We will look at each of them separately.

Table 2.6 shows the discriminators used by the different schemes.

Table 2.6 *Discriminators used in the area of making observations and measurements*

LEAG "A"	
Observational skills	Amount of prompting/help, accuracy and precision of observations, complexity of scales
LEAG "B"	
Observational skills	Complexity or difficulty of observations
Measurement skills	Complexity of scales, number of variables, selection of most suitable instrument
MEG	
Measurement	Accuracy
Observation	Accuracy, relevance, detail and systematic approach
NEA	
Making quantitative observations	Complexity of scale
SEG	
Making observations	Range of observations, amount of care and accuracy

Qualitative observation

Qualitative observation can involve recognising differences and similarities between different objects or phenomena or the observation of changes or transient events.

In Physics, differences or similarities are usually restricted to one or two properties, for example, the brightness of different lamps or the pitch and loudness of different sounds. This contrasts with the situation in Biology where differences or similarities between organisms can apply to a large number of different characteristics.

One problem which besets observation is that of relevance. Objective 3.2.5 requires pupils to be able to extract data relevant to a particular context, and some of the discriminators involve the relevance of the observations. Relevance depends upon the framework from within which the observation is made. At this level we need to be sure that the outcomes of observational exercises do not depend on prior knowledge.

Some direction will be necessary in the instructions provided. For example, in an experiment involving the effect of a length of resistance wire in series with a lamp, one would expect a reference to the brightness or appearance of the lamp in the instructions. Thus:

"How does the lamp change as the length of wire in the circuit is increased?", would be preferable to

"Alter the length of resistance wire in the circuit. What happens?"

In the observation of changes, the complexity of the problem depends on the number of factors which are changing, as well as the degree and rate of change. It is easier to observe situations in which only one property changes, or in which only one property is changing at any particular time.

There are, then, two different factors which we need to bear in mind when designing observation exercises. The complexity and/or the difficulty of the exercise may be the discriminator (as in the LEAG syllabus B) and this will be defined by the syllabus.

Alternatively, the level of performance may depend on what we might term the quality of the observation.

The three points which we would look for are:

Accuracy:	Is the description (written or verbal) correct?
Completeness:	Does the description include all of the factors required?
Relevance:	Are the factors reported those which were required by the problem as it was set? Is the inclusion of any additional observations supported by sensible reasons?

Measurement

The ability to measure includes a very wide range of skills and processes, some of which are manipulative. These include:

- selection of a suitable instrument,
- precautions to ensure reliability of measurement – not spilling or splashing materials, correct positioning of thermometers etc,
- allowing for zero errors and avoiding parallax error,
- accurate reading of the scale and assigning magnitudes and units to readings, and
- taking precautions to ensure reliability and accuracy by repeating measurements.

For many situations, the sorts of problems which affect qualitative observation do not apply. In most exercises which involve measurement, the instructions would involve a statement of the quantities to be measured. It is only in the area of experimental design that candidates might have problems of relevance.

The discriminators fall into two main groups, the ability to read scales and the techniques required to ensure accurate measurement.

There are a number of factors which can affect performance in reading scales, some of which depend on the nature of the exercise and some of which depend on the form of the scale; these can be used to discriminate between performance levels and also as a means of differentiating exercises.

Work reported by the APU suggests that the level of performance is higher when the pupil is actively involved in the measurement task. Thus, candidates perform better when asked to measure out a given volume of liquid than they do when they are asked to read a measuring cylinder containing a volume of liquid. It is likely that observing the movement of a pointer or surface towards a scale marking provides some help. This is especially true when inverted scales, such as forcemeters, are involved. If this be the case, then there are implications for stations or circus-type exercises which we will consider in the next chapter. The other factor which makes reading scales difficult is that the pointer may lie between graduations so that interpolation or rounding is necessary.

The simplest form of scale is a linear scale, having every division and sub-division numbered and the units marked on the scale. There are very few such scales in the majority of school laboratories. Most scales do not have the sub-divisions numbered. The scale becomes more difficult to read when the sub-divisions are not unit sub-multiples of the numbered graduations and this is used as a discriminator in the NEA syllabus.

The ability to read both scales with "unit" subdivisions and non-unit sub-divisions to at least the nearest scale division should be within the capabilities of the majority of pupils at GCSE level.

At a higher level, multi-scale or multi-range instruments or a more complex scale, non-linear and reversed (eg a resistance meter) are specified in the LEAG syllabus B. Obviously, one would have to set exercises which used such devices where the syllabus required, but as a general rule their use should be avoided for all but the most able pupils.

As a first step towards designing exercises it might be useful to draw up a table of the instruments available and the types of scales which they employ. Such a table of instruments from the author's department is shown in Table 2.7. There are a few of these instruments commonly available in class sets or half-class sets in most centres.

Table 2.7   Some instruments with linear scales which can be used for the assessment of the ability to make measurements

Quantity	Instrument	Unity scale	Non-unity scale
Mass	lever arm balace	*	
Length	metre rule	*	
Volume	measuring cylinder	*	*
Angle	protractor	*	
Time	stopwatch	*	*
	stopclock	*	
Force	newtonmeter	*	*
Pressure difference	manometer	*	
Pressure	Bourdon gauge	*	
Temperature	thermometer	*	*
Voltage	voltmeter	*	*
Current	ammeter	*	*
Current	milliammeter	*	*

Given such considerations, we can now list those aspects of performance which might be used for assessment.

Selection of measuring instrument
Is the instrument selected capable of measuring the correct quantity?
Are the range and sensitivity of the instrument appropriate to the measurement to be made?
Is the instrument selected liable to be put at risk if it is used as the candidate intends?

Performance of the measuring operation
Are all reasonable precautions taken to ensure that the measurement will be valid?
Are measurements repeated or checked?

Reading scales
Does the candidate recognise and allow for zero errors?
Are the readings made with due regard for parallax?
Is the scale reading translated into the correct magnitude and are the correct units assigned to it?
Is the reading correct to within one scale division (or interpolated if appropriate)?

Handling experimental data
(Objectives 3.2.1, 3.2.3, 3.2.4, 3.2.5, 3.2.6, 3.2.7, 3.2.8, 3.2.9, 3.2.14 and 3.2.15 in Science)

The objectives concerned with this area fall into three groups: recording and reporting, processing data, and drawing conclusions. The discriminators adopted in the various skill areas are shown in Table 2.8.

Table 2.8   Discriminators used in the area of handling experimental data

LEAG "A" **Recording and reporting**	Format for presenting data, amount of help in reaching conclusion, completeness of report
LEAG "B" **Use and analysis of data**	Complexity of problem analysed
Recording and reporting	Format for presenting data, logical presentation, completeness of report
MEG **Handling experimental data**	Amount of help in selecting format/reaching conclusion, conclusions qualitative or quantitative
NEA **Present results**	Completeness and ease of access to information
Draw conclusions	Amount of help and depth of conclusion
SEG **Recording and processing**	Adequacy of record, processing and conclusion

Recording observations and measurements

The criteria which define performance levels in recording observations are similar to those for making observations. The difficulty, as far as the assessor is concerned, is separating the two processes. What might, at first, seem an inadequate record might well be a comprehensive record of inadequate observations! The teacher must determine the extent and quality of the observations from discussions with the candidate before attempting to assess the record.

It would seem obvious that an attempt to assess this area would be coupled with the assessment of observational skills so that the teacher would already have ascertained the relevance of the observations.

The points which might be used for assessment are:

Accuracy
Is the written report a correct description of the observations made by the candidate?

Completeness
Does the written report include all of the observations made by the candidate?

Method of reporting
Does the report make use of a variety of methods of conveying information (written and diagrammatic)?
Is the method adopted the most appropriate? Is the record systematic?

The recording of measurements is concerned with the presentation of quantitative data in an appropriate format. The APU classifies the formats in four styles: random, prose, ordered and tabulated. The APU publication, *Science in Schools: Ages 13 and 15, Report No. 3*, explains these terms more fully than this summary does.

Prose and ordered presentations do not allow easy access to the data and would therefore be considered to indicate a lower level of performance.

For our purposes, the most appropriate format is a table. A properly constructed table should include data presented in columns with headings and units included. Furthermore, at the highest level of performance, one would expect that the table would include the results of all measurements and not just those quantities which are derived from the measurements. This enables us to check whether any inconsistency is the result of arithmetic error.

The three levels of performance are:

At the **lowest level** the information would be recorded as a prose account, usually as a sequence of statements. The candidate would be able to record data in a pre-prepared table.

At the **intermediate level** the candidate might record data as an ordered set of statements or in a table which might omit units or some data.

At the **highest level** the candidate would present results in neat tables, including all headings and units, with all measurements recorded as well as derived quantities.

Reporting (Objective 3.2.3 in Science)

Only the LEAG syllabi include assessment of a written report which includes a description of procedures. As in the case of reports of observations, the criteria by which we would assess the report would be accuracy and completeness, but we must also concern ourselves with the clarity of the report and a logical presentation.

Processing and analysing data

There are two processes which feature in this area, graph plotting and drawing conclusions.

Performance in graph plotting can be split into a number of stages:

- Selection of axes and suitable scales.
- Labelling axes correctly, including units.
- Plotting points accurately.
- Drawing the best curve (including neatness)

The three levels of performance are:

At the **lowest level** candidates might produce bar charts or be able to draw graphs on pre-prepared axes. They would be able to plot points for whole number values but would experience difficulty with decimal values.

At the **intermediate level**, candidates would be able to select their own scales and axes, possibly with help, and to plot points accurately.

At the **highest level**, candidates could plot graphs correctly without assistance, selecting and labelling suitable scales and axes and drawing a straight line or neat curve which represents the best fit to the plotted points.

The ease with which we can analyse performance in graph plotting contrasts with the complexity of the problems which beset the remainder of the area. The large number of objectives associated with this area is a reflection of the wide range of different processes which might be involved.

Within the context of traditional experimental work, our first concern might well be the drawing of conclusions based upon experimental data. There are a number of different processes which we might classify as conclusions.

At the simplest level this would involve classification of observations and recognition of similarities or differences.

At an intermediate level, candidates could perform a simple calculation of a derived quantity or recognise a simple pattern or trend in the relationship between two variables.

At a higher level we would expect that relationships were expressed more exactly either in verbal or mathematical terms. At GCSE level it is likely that one would consider making assessments only in situations where relationships are linear, although the LEAG syllabus expects that at the highest skill level candidates should be able to identify inverse proportion and square laws.

In addition to drawing conclusions we would also expect that candidates should be able to evaluate their data critically and to recognise limitations and areas of uncertainty or unreliability in their experiments.

Designing and carrying out experiments
(Objectives 3.2.9, 3.2.10, and 3.2.13 in Science)

Of all of the abilities which are to be assessed, it is this area which causes teachers most concern because for most of them it has not been included in previous physics courses at this level. Indeed, it has not featured in the majority of courses at A-level either!

All of the schemes have included the ability to design experiments in order to solve problems or test hypotheses as an area for assessment. The MEG syllabus also includes experiments to measure derived quantities which is a more straight-forward aspect of experimental design.

There are two aspects to the problem. The first is concerned with the types of problem which might be suitable for exercises and the second with what might be an acceptable performance at this level.

Most experience of investigations and problem solving approaches at this level has involved extended projects in the more applied areas such as electronics and technology. Although the opportunity for pupils to engage in extended practical projects in physics might be seen as desirable, it is clearly not a practical proposition as a vehicle for assessment for the majority of centres. A more detailed discussion of the types of problem which might be set will be given in the next chapter.

The other concern is that of acceptable performance. It is important that we do not expect too much from candidates. A rule-of-thumb guide is that the planning and execution of the exercise should be capable of being completed satisfactorily in about half an hour. This is the approximate period needed by pupils to complete the tests in the APU testing.

Unfortunately, the performance descriptions provided by the published schemes are of little assistance either. The discriminators used are shown in Table 2.9.

Table 2.9   Discriminators used in the area of design, planning and execution of experiments

LEAG "A" **Design, planning and execution**	Amount of prompting/help
LEAG "B" **Designing and planning experiments**	Amount of progress towards a reasoned solution to problem
MEG **Design and execution of experiments**	Method of approach to problem, amount of help in translating plan into action
NEA **Design and carry out experiment**	Amount of help
SEG **Designing the experiment**	Success in identifying variables and selection of procedure

Most of the schemes rely on the amount of help to discriminate between the performance levels, which does not provide the teacher new to this area with much of a guide as to how to design suitable exercises. It also introduces the problem of collusion which has been discussed above. The problem of collusion can be minimised by using a range of exercises so that candidates will be unable to obtain clues from their neighbours, but this approach compounds the problem of developing suitable exercises! There are ways of reducing this problem however, and some of these will be considered later.

As a starting point we will look at the skills and processes involved in the area. Many of these have been discussed above, and so we will concentrate on those aspects which are peculiar to experimental design.

The unique aspect of this skill area is that of recognising the variables involved in a problem, and of designing an experiment which allows investigation of the relevant variables and controls the remainder. Such problems as the manipulation of apparatus, the range and accuracy of measurements or observations, the recording and analysis of results and reaching conclusions are all assessed in the other skill areas. They do, of course, have an important part to play in the successful solution to the problem, but it is important that we do not place undue emphasis on them if they are being assessed elsewhere.

Our main consideration should be the planning stage where the pupils decide on the relevant factors and produce ideas for solving the problem and on the evaluation stage when they examine the outcome of their experiment critically. With this in mind we can draw up outlines of the levels of performance as follows.

At the **lowest performance level**, candidates work empirically with little thought to an overall strategy or planning. They will have difficulty in analysing the problem in terms of an experimental procedure without considerable assistance.

At the **intermediate level** we could expect a recognition of the variables involved, but candidates would need some assistance as to which variables to control. The planning involved in the experimental work will involve a selection of suitable apparatus and a workable sequence of operations, although candidates may require occasional help.

At the **highest level** we would expect candidates to analyse the problem in terms of the important variables and recognise those which must be controlled, to plan the experimental work in a sensible and logical manner, and to modify their strategy when appropriate. In addition, the most able candidates might well be able to suggest alternative strategies and to make reasoned decisions as to the most appropriate.

Summary

In this chapter we have considered how the objectives from the National Criteria have been translated into schemes for coursework assessment by the various examining groups. These schemes have provided descriptions of performance at different levels, but in general they have not given guidance as to how these descriptions might be translated into an operational scheme of assessment exercises. As a first stage in this process we have considered the discriminators which enable the assessor to distinguish between different levels of performance, and we have provided a number of guidelines which should be considered when designing exercises.

The general nature of the analysis will require teachers to make a selection of those aspects of performance which fit the particular syllabus adopted. Having made such a selection there is the further consideration of the organisational problems involved in making assessments. The next chapter will consider possible strategies for managing assessment sessions.

3 Strategies for managing assessment

Introduction

"So what's the problem? I have been teaching practical skills for as long as I've been teaching physics. I've been making decisions about levels of pupil performance so that I knew who could do it and who couldn't."

Many teachers could make the statements above quite truthfully and could then go on to provide their own list of answers to the question. The problem is twofold. Firstly, assessing to criteria, instead of making subjective estimates of performance, increases the amount of time spent on each assessment and the planning required in the design of exercises. Secondly, we are now required to make assessments of every pupil on at least the number of occasions specified by the syllabus.

The sheer quantity of assessment means that for many teachers, the problem is one of management. Class sizes of 30 which are equipped for working in groups of four, inadequate technician support and lack of any additional time for the extra work involved, will be familiar concerns to most.

This chapter describes a number of strategies which are suitable for managing internally assessed practical work. In some cases special problems are discussed and the relevance of each strategy to the various skill areas outlined in the previous chapter is considered.

Where appropriate, the strategies are illustrated with examples of exercises and further examples are provided in the specimen exercises.

Written tests

A number of the skill areas to be assessed involve process skills which, although they are associated with experimental work, are not practical skills as such. These include the presentation of results, drawing conclusions and the design and planning stages of an investigation.

The National Criteria require that only 10% of the total marks be awarded on the basis of experimental and observational work in the laboratory. This means that up to half of the marks available for coursework assessment could be awarded on the basis of pencil and paper exercises. Only the NEA syllabus specifically states that the use of written exercises is an acceptable method of assessment for some skill areas. The majority of the other schemes make no mention of the approach and so, presumably, it would be permissible. The exception is, of course, the SEG syllabi which use individual assignments to assess all skill areas. It would not be possible to use written exercises in this situation.

Written exercises have many advantages from an organisational point of view. They can be administered to a large number of pupils under examination conditions, and they make no demands on apparatus or laboratory accommodation. Further, it is easy to isolate the skill area to be assessed using written exercises. All pupils are provided with the same information so that we can be certain that at the least, they have started

from the same point. Not least, marking can be carried out at a convenient time and a permanent record of performance is produced.

On the other hand, the exclusive use of written exercises does not develop an holistic view of experimental skills. More importantly, from the point of view of assessment, they may not be a true measure of a candidate's ability. The APU's findings suggest that there is a difference in performance levels, between pencil and paper exercises and tests involving practical versions of the experiments, in some skill areas but not in others (APU, 1985). These results suggest that we must be careful in the use of written exercises.

Areas for which the use of written exercises might be appropriate are:

- recording and processing results,
- analysis of data and drawing conclusions
- experimental design.

Written exercises could be used to reinforce the acquisition of skills in most skill areas. Many of the so-called written practical papers which have been used by examination boards in recent years can provide examples and a further range is included in Table 3.1 on page 12.

A supply of such exercises would be a useful addition to the department's range of teaching materials. Unfortunately, in many cases they tend to be so apparatus specific that there is little alternative but for each centre to develop such materials to suit the apparatus available. Indeed, this is an objection which has frequently been voiced over the use of written papers to test practical skills.

Assessment within normal practical work

It might be argued that since existing courses involve significant amounts of practical work, then it might reduce the extra workload if candidates could be assessed during the course of such work. A checklist of skills involved in a particular experiment, laid out on a grid with candidate names along the other axis, could be used by the teacher to record the performance of individuals as the opportunity arose. The principles involved in constructing such grids are discussed in the next chapter.

The teacher would not set out to test achievement in all areas for each candidate. Indeed, it is unlikely that observation of a single skill for each candidate would be possible. A simple calculation of the time required to assess all pupils in a group shows just how unlikely!

The marks representing the performance in a skill area for a particular candidate are entered on the grid as the assessment is made. The missing skills and candidates who are absent from this grid would be assessed on other occasions.

In the majority of experiments it would not be possible to assess the complete range of skill areas required by the syllabus. In such cases an extension of the traditional experiment, so that it involves problem solving or hypothesis testing abilities, can be envisaged. Sample exercise 12 illustrates how a traditional

Table 3.1   Examples of questions which could be used in written tests

(i) Manipulation of apparatus

Which of the following statements about the setting up of a cathode ray oscilloscope is/are correct?

A The trace should be adjusted for maximum brightness.

B When making accurate measurements of voltage, the Y-gain or volts/cm control should be set to a level which gives as large a deflection of the spot on the screen as possible.

C The time base should be set as fast as possible when studying alternating voltages.

D The most accurate way to measure the peak to peak voltage is with the time base switched off.

In the circuit on the right, a power supply is used in an experiment to find the resistance of a 6 V lamp.

Put the operations listed below into a sequence which could be followed so that none of the apparatus used will be damaged.

A Switch on the power supply

B Make sure that a fuse is properly fitted in the power supply

C Set the volts control to 6 V d.c.

D Connect up the circuit, checking that the meters are connected with their positive terminals as shown in the diagram

E Plug in the power supply

F Check that the voltmeter and the ammeter are in the correct positions

G Follow the circuit round carefully to make sure that there are no short circuits

H Check that the connections to the power supply are in the correct terminals and that the "volts" control is set at not more than 6 V.

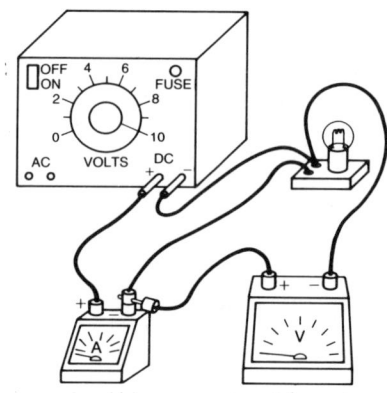

(ii) Making quantitatative measurements

The following describes an experiment to measure the volume of a glass marble.
A measuring cylinder was partly filled with water. The level of the water is shown on the left below.

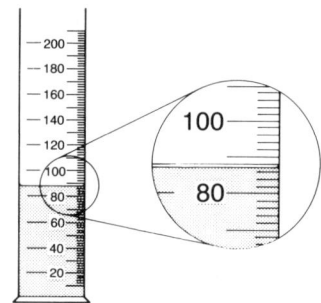

The volume of water is _____

Ten glass marbles were then added to the cylinder and the new level is shown on the right.

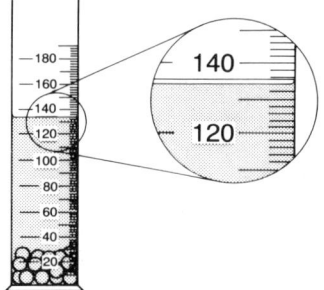

The volume of water plus ten marbles is _____

The volume of ten marbles is _____

The volume of one marble is _____

(iii) Handling experimental data

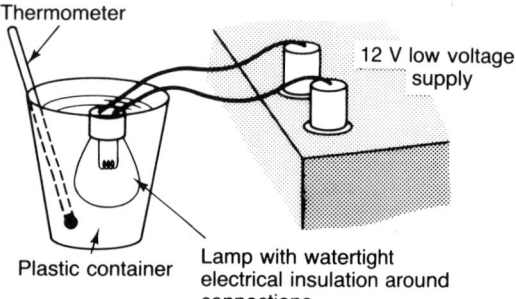

Thermometer

12 V low voltage supply

Plastic container

Lamp with watertight electrical insulation around connections

A 12 V, 3 A lamp bulb is used as an immersion heater by placing it in a thin plastic container containing water.

The temperature of the water is measured every 4 minutes and the results are shown below.

Temp. (in degrees C)	12	13	14	15	16	17	18	18	18
Time (in minutes)	0	4	8	12	16	20	24	28	32

(i) Plot a graph of temperature against time.

(ii) Explain why the temperature stops rising when the water reaches 18 degrees (the lamp was not switched off).

(iv) Planning investigations

Daniel says that if you wrap ice cream in newspaper it helps to keep it frozen. Alice says that if you soak the newspaper in water before wrapping the ice cream it stays frozen even longer. Design an experiment that you could do to test the claims of Daniel and Alice.

experiment on density can be modified to extend the range of skills involved. Table 3.2 shows a sample marking scheme for this exercise based on the SEG scheme.

Table 3.2   A mark scheme for sample exercise 12 using the SEG syllabus

Criteria	Mark (1 or 0)
A1 Measurements of mass and volume	
A2 Selection of most appropriate methods of measurement for each sample	
B1 Apparatus for measuring mass and volume identified	
B2 Apparatus used correctly without major assistance	
B3 Apparatus used correctly unaided	
C1 Some measurements obtained	
C2 All necessary measurements of mass and volume obtained	
C3 Careful technique with all precautions	
C4 Equipment used accurately	
D1 Observations recorded	
D2 Densities calculated	
D3 Samples identified	

For schemes which allow assessment of individual skill areas, the exercise could be carried out by pupils working individually, in groups, or a combination of the two. This would allow assessments of a number of candidates to be made without too great a demand on apparatus or accommodation. If assessments are to be made on those working individually, one would obviously need to decide at the beginning of the lesson which pupils are to be assessed.

Alternatively, the teacher might split groups for part of the exercise to allow the assessment of an individual's performance in one skill area, and then allow the group to work together to complete the exercise. This strategy will require that exercises are staged so that there are suitable points at which the groups can split or join up. Worksheets will require check points at which pupils draw the teacher's attention to the stage reached so that the appropriate action can be taken.

By careful design of the exercise, it might be possible to assess performance by referring to the written records produced by the candidate. This would involve practical lessons being conducted under examination conditions to ensure that the results were solely the work of the candidate. Whilst such a procedure has the advantage of increasing the number of assessments which can be made on a given exercise, it would involve a major change in the character of practical sessions which few would see as desirable.

Formal practical tests

The problems which might arise if assessment were carried out in the course of normal practical sessions, such as coping with problems when apparatus does not function, interruptions from outside of the class, giving due regard to safety, helping pupils who have difficulties with understanding or following instructions and all of the other events which punctuate science lessons, are the spectres which inhabit the nightmares of teachers contemplating GCSE.

The National Criteria allow that coursework can be assessed periodically or continuously during a part or the whole of the course. Further, a maximum of 20% of the total marks can be assessed by means of formal practical tests. It would, therefore, seem permissible to assess coursework by a series of practical tests administered throughout the course or during the period stipulated by the particular syllabus.

At first sight, formal tests might seem the most efficient method of assessing many of the skill areas outlined in the previous chapter, as well as the only way of avoiding the problems listed above. Indeed, a number of existing courses use the technique of teacher assessed practical exercises during the course as one component of the examination. There is, however, a significant increase in both the level of teacher involvement and the detail required by the assessment criteria for GCSE over these existing schemes.

There are a number of drawbacks to the exclusive use of formal tests, some of which are organisational in nature and some of which conflict with the spirit of GCSE. The more important of these are:

(i) Group size

Given the size of teaching groups in some centres it would be possible to assess only half groups on any one occasion. If the number of skills which are to be assessed by an exercise be limited to one or two (and there are sound educational reasons for doing this which are discussed below), then it follows that a total of about twenty hours would be needed for carrying out assessment. For half of this time part of the group would require supervision and accommodation. Bearing in mind the fact that this situation is liable to be repeated for further groups in physics and in the other sciences, it becomes obvious that the administrative problems posed would be intolerable.

(ii) Absentees

When fixed times are set aside for assessment and the number of assessment occasions is relatively large, the problem of absentees can become significant. It seems improbable that further special provision could be made on top of the disruption caused by the normal assessment programme.

(iii) Apparatus and accommodation

If tests are to be carried out under examination conditions there must be a reasonably large space around each candidate if they are carrying out the same exercise. The use of a set of identical exercises could put a strain on the provision of a sufficient quantity of apparatus. A further consideration is that apparatus which is being used for assessment will not be available for normal practical work. In schools which timetable a number of groups at the same time (so that sixty or more hours of the course might involve apparatus being in use for assessment of one group or another), the opportunities for developing skills will be severely restricted.

(iv) Appropriate assessment

The National Criteria require that the assignments for assessment must be designed to make appropriate demands on all students in a teaching group. The administering of formal tests to a group would tend to penalise those members of the group who require a longer time to develop skills. In addition, a test which employs the same exercise to assess all candidates would be unlikely to make "appropriate demands" on all members of the group.

These criticisms would suggest that in the context of GCSE, formal testing alone will not meet the needs of the teacher or the requirements of the National Criteria. There is undoubtedly a place for formal tests, for instance in assessing skills such as translating information or drawing conclusions (which can be carried out as written exercises), but they must be seen as a part of the armoury to be employed when designing an assessment scheme.

Table 3.3   Example of an exercise which could be used in a formal practical test

(i) Manipulation of apparatus/following instructions

Assemble the apparatus provided to produce the circuit shown in the diagram. When you have completed your circuit ask your teacher to check it.
Candidates could be asked to make measurements using the completed circuit, and this could be used for further assessment at the discretion of the teacher.

(ii) Making quantitative measurements

The apparatus shown in the diagram below is assembled for you. Each resistor has a resistance of 3.9 ohms. Complete the circuit by connecting the flying lead (labelled F) to the end terminal on the row of cells. Measure the current flowing through the ammeter. Move the end of the flying lead from terminal A to terminal B. Measure the new current. Measure the current flowing through the circuit with the flying lead connected to terminals C and D in turn.

Record your results in a table.

Candidates who obtain a set of results can continue with the rest of the exercise to provide the opportunity to be assessed on handling data and drawing conclusions.

Plot a graph of current against the number of resistors in the circuit.
The total resistance in the circuit is given by the equation

(total resistance) = (number of resistors) × 3.9 ohms

What does your graph tell you about the connection between the current and the resistance in the circuit?

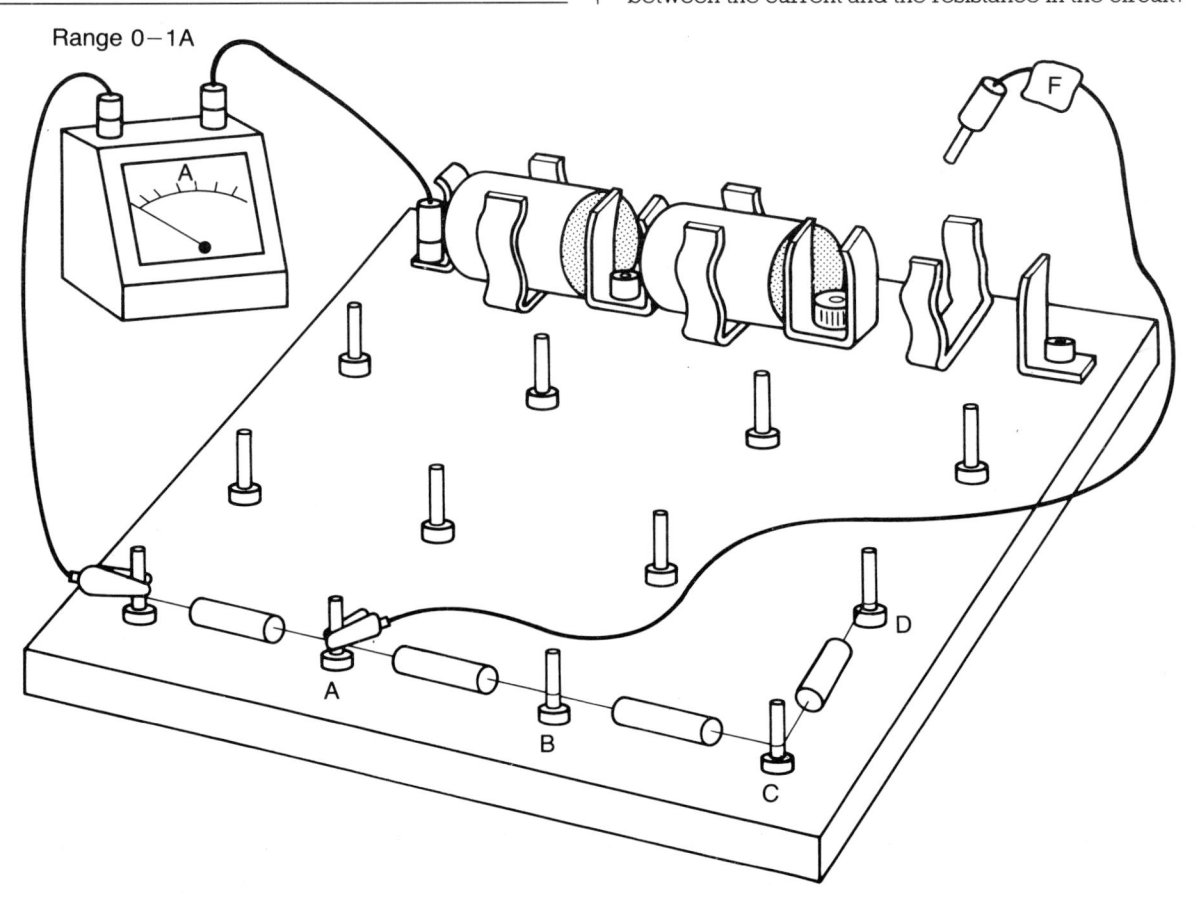

There is however one skill area in which some type of formal test situation is the only possible approach. This is the area of designing and carrying out experiments. There is no alternative but to set up special tests with part groups for the assessment of this area. Problems of space and equipment will limit the maximum group size, even if we accept that the pupil's written account will provide sufficient information for the major part of the assessment. I would suggest that it would not be possible to assess more than ten or twelve candidates effectively in one session although colleagues report that they have managed to assess groups of 15 successfully.

Demonstrations

As science teaching has become more pupil centred, there has been a tendency to minimise the role of the class demonstration. Although one would not like to see a wholesale return to demonstrations, there are situations where they can provide a suitable vehicle for assessment.

A demonstration places less demand on resources and technician time than does a class experiment, but more importantly, in the right context it has a valid role to play in assessment.

Demonstrations can provide pupils with common data for recording, processing and deriving conclusions. The management of the class would need to be such that pupils could not discuss their work whilst the assessment tasks were being carried out.

The demonstration also has the advantage of isolating the assessment exercise from the complexities of the rest of the experiment. Less able candidates can be presented with tasks which are appropriate to their abilities even in the context of quite difficult experiments, and areas of the syllabus which are not suitable for pupil work at this level (such as radioactivity) can still be used for assessment.

Stations or circus exercises

It might be argued that, since the majority of the assessment schemes have adopted an "atomistic" approach to practical skills, then, the most appropriate format for assessment exercises would be one which mirrors this approach. In other words, that the best way to assess individual skills is by means of individual exercises for each skill.

One way of organising such a set of exercises would be to set up a number of "stations" around the laboratory. Pupils would move from one station to the next performing each exercise. It is a technique which is widely used in lower school science lessons and A-level practical examining because it enables a wide range of experiences, or tests, to be performed in a relatively short time.

The approach could be used to test a number of different skill areas, or to provide a number of opportunities to demonstrate performance of the same skill in different situations. It is not necessary for every pupil to carry out every exercise, but it is important that the time required for each station is the same.

The advantages of stations are offset when the need for direct observation by the assessor limits the number of skills and/or pupils which can be assessed on any one occasion. Careful design of the exercises so that they produce an "end-check" (a written record which can be marked later) could minimise this problem. Furthermore, for some types of exercise, the space required for each station will limit the number of candidates who can be assessed in a session.

We must, therefore, restrict the assessment to part of a group, and possibly use a second assessor, if the session is to run smoothly. A further consideration which will limit the number of stations, is the need to check apparatus after each pupil's use. It is important that the teacher be not overambitious!

One problem which might well arise is that of collusion. Practical sessions with a large number of pupils are not like formal examinations. The teacher cannot stand at the front of the laboratory observing the whole group to ensure that every individual's work is just that! Indeed, it might well be important that the teacher pays particular attention to one or two exercises for the purposes of assessment.

In order to minimise the problem, it might be possible to design exercises so that each pupil achieves a unique result which is capable of being checked before or after the assessment session. For example, task D in specimen exercise 11 could be modified by altering the values of the resistors between candidates. The teacher would require a record of the particular value for each candidate. Numbered or code lettered boxes can help with this problem.

Table 3.4   Examples of exercises which can be used as stations

---

(i) Manipulation of apparatus

1 Connect the ray box lamp to the power supply.
2 Place the lens on the lens holder.
3 Set the lens holder so that there is a distance of 20 cm between the mark on the ray lamp base and the mark on the lens holder.
4 Place the screen so that there is a sharply focused image of the lamp filament on the screen.

WHEN YOU HAVE FINISHED, LEAVE THE APPARATUS SET AS YOU HAVE ADJUSTED IT AND THEN ASK YOUR TEACHER TO SEE WHAT YOU HAVE DONE.

(ii) Making qualitative observations

1 Hang each of the containers from the springs and observe the results.
2 Describe your observations.
3 Set each container oscillating – pull it downwards by about 5 cm and then let it go.
4 Observe the oscillations of the containers and describe your observations as carefully as possible.

The station consists of three expandable springs suspended from clamps, and three identical containers with different masses of between 200 g and 700 g. There should be some means of hanging each container from the springs.

These stations could be modified to enable the assessment of making quantitative observations, recording and handling data and drawing conclusions.

New ways of providing data

The assessment of individual skill areas provides us with an opportunity to use new ways of presenting data to pupils. Devices such as micro computers, data-loggers (like VELA), video and audio cassette recorders, can all be used.

Micro computers can be used to generate data which can be output on the screen for processing by the pupil, or through a digital to analogue convertor to a suitable meter for assessment of measurement skills. Even when teachers are not keen on using such a procedure for assessment, it can be used to develop the skill of reading scales. The analogue output of the VELA can be used in a similar way.

Alternatively, simulation programmes may be used to provide data for processing, although programmes which produce representations of scales on the display would not be suitable for assessing measurement skills.

Records of experiments can be presented visually using a VCR. Schools' broadcasts sometimes include sequences in which experimental results are presented. These can have the advantage that data from experiments which are dangerous, or which involve expensive apparatus, can be made available.

Some programmes have involved materials from industrial and other external sources which could be used as a basis for assessment exercises. This provides an opportunity to extend the process skills into the social and technological areas which are an important part of GCSE syllabi. The teacher will need to keep a lookout for suitable programmes.

An alternative for schools (or groups of schools) which have access to a video camera, might be to record some standard experiments carried out by sixth-formers or teachers. However, unless those involved are familiar with the techniques of video recording, careful planning and editing will be necessary to ensure that the result is unambiguous and easy to follow. A further problem is that such exercises might tend to become ossified as the effort of making them discourages frequent revision or replacement.

Even audio recordings can provide a source of data for recording and processing. There is, however, a need for careful editing of recordings and an element of caution in interpreting the results. The dialogue which accompanies a group carrying out the experiment, may be easy to follow for those engaged in the process, accompanied as it is by practical manipulation and visual signs. The script below (which could also be used as a written exercise) is similar in form to a specimen question from the NEA Dual Award Science Syllabus. It shows how confusing a verbatim record might be in the absence of the other information which would be available to candidates carrying out a full practical exercise.

A group of pupils were investigating how the current through a light dependent resistor (LDR) varied with the brightness of light shining on it. They put it close to a car headlamp bulb and measured the current flowing through the bulb and the potential difference across it. They covered the bulb and the LDR so that the light from the room would not affect their measurements. The apparatus they used is shown below.

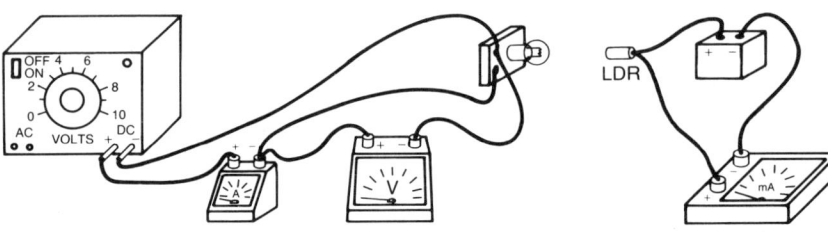

A record of what was said during the experiment is given below.

Albert: Before we switch on the power supply. What are the readings on all the meters?

Beryl: The ammeter and the voltmeter both read zero.

Satnam: And there is no reading on the milliammeter.

A: I've switched on the power supply at its lowest setting.

B: The voltmeter reads point eight and the current through the lamp is point six of an amp.

S: The current through the LDR is tiny! It's only point zero two milliamps.

A: I'll increase the supply and you two shout out your readings.

B: Two point three volts and one point one amps.

S: The current has gone up to nought point seven amps, no I mean milliamps.

B: Four point three volts and the lamp current is one point five amps.

S: Wait a minute. I have to change the range on the milliammeter. That's better. Three point four milliamps.

B: Now the voltage across the bulb is six point one and the current through it is one point eight amps.

S: The LDR current is now eight milliamps exactly.

A: Another two readings should be enough. What are the readings now?

B: Exactly eight volts and its a bit more than two point one amps. Let's call it two point one.

S: Hang on! The pointer has gone off the end of the scale. I'll have to change the meter. The current through the resistor is thirteen milliamps.

A: Last reading.

B: Ten volts and a current two point four amps.

S: And the current is eighteen milliamps.

(i) Draw a table of the results of the experiment.

(ii) Use the equation

(power) = (potential difference) × (current)

to work out the power supplied to the bulb for each setting of the power supply, and include your answers in an extra column in your table.

(iii) Plot a graph of current through the LDR against power supplied to the headlamp bulb.

A script such as this could be set as a written exercise or as an audio recording. The problem with the latter approach is that pupils will probably find it difficult to follow the account if it is just spoken. A more suitable strategy would be to provide the script and the audio account which would also help pupils with reading difficulties.

The problem might also be made easier by providing candidates with a copy of detailed instructions for the experiment and allowing them to see the apparatus set up for it. Even then, it is still a daunting task trying to sort out the important points from the dialogue.

These new techniques are not a substitute for practical work by the pupil, but they can provide additional opportunities for developing skills. Because they have outcomes which do not require direct observation by the teacher, they may be useful as vehicles for assessment when a number of pupils need additional assessments because of absence, or towards the end of the course when time is short.

Developing skills

Although not strictly a strategy for managing assessment, careful attention to the development of skills in pupils is the most effective strategy which is available to us. The actual process of assessment will be much easier in groups where pupils are properly prepared for practical work. Furthermore, as was mentioned in the first chapter, our principle concern is the teaching and development of skills rather than their assessment.

For the majority of skill areas there will be plenty of opportunity for the development and practice of skills during normal laboratory and classwork. The area of experimental design for problem solving, however, presents us with difficulties. Because of this it will be considered in detail.

Exercises for design can be of three types:

1 Experiments which provide an answer to a specific question or which require a specific measurement to be made. Examples are "Design an experiment to find the maximum current which a 3-amp fuse can carry", or, "Design an experiment to measure the biggest load which can be supported by each of the magnets provided".
2 Experiments to test a hypothesis or the validity of a statement. "The time for one complete swing of a pendulum is not affected by the mass of the bob" is an example of one such statement.
3 Open ended investigations such as "Investigate the factors which affect the performance of a paper aeroplane", or "What is the best material for a non-slip floor?".

As a general guide the first consideration should be to keep the exercises simple. Restricting the experiment to a limited problem or giving clues to the types of measurements required, as in 1 and 2 above, are ways of simplifying problems.

Open ended problems often involve more complex situations. Problems such as "Which is the best heater?" or "Which is the best insulator?", involve judgements which are not entirely physics based. The "best heater" may be the most robust, or the cheapest, or the most efficient (to cite three interpretations). We should try to avoid such ambiguities when designing the problem and if they become apparent from the candidate's answer then the marks should be awarded for success

in arriving at a solution to the candidate's perception of the problem.

Another method of reducing difficulty, as well as easing the problem of organisation once the session is under way, is to provide some of the apparatus, or a suitable range of apparatus, plus some unsuitable items. For some exercises, the provision of a number of test pieces can be used to restrict the range of variables involved in an experiment without making it too easy for candidates. Sample exercise 13b illustrates this technique.

APU tests of problem-solving skills, provided pupils with a range of apparatus from which they could select the most suitable items for their adopted strategy. The range and types of apparatus provided might well point the direction which the investigation might take. The opportunity to handle the apparatus and to perform rough trials before making a final decision on the strategy, is a more satisfactory approach (and a more "scientific way of working") than asking for a written plan without the opportunity for trials.

The characteristics, unique to this skill area, the need for individual work over an extended period of time and the open ended nature of the work, might well lead to concern over the time needed to develop the skills prior to assessment. The assessment itself will cause disruption and it would be undesirable to cause further disruption by setting aside special "trial" sessions to provide pupils with opportunities to practise.

APU experience suggests that pupils of a wide range of abilities are surprisingly good at designing experiments to solve problems – even without special training. This indicates that teachers' fears might well be groundless, or at least unnecessarily acute.

The skills involved can be developed in the course of normal laboratory work if the approach to experimental work be sympathetic to the design and problem solving approach.

Pupils should not be provided with pre-packed prescriptions or recipes for carrying out experiments. Instead, the teacher should adopt a questioning approach to setting up the problem involved in an experiment, and then in the design of the practical work to achieve a solution. The term "problem" is used here in a very broad sense. An investigation of the current-voltage relationship for a filament lamp is just as much of a problem, when approached in the right way, as is an investigation into the bounciness of tennis balls.

An example of the approach is given below.

Teacher: You have found out that metals are conductors of electricity. When you put a piece of metal wire into a circuit with a bulb and a battery, the bulb lights. What makes the bulb light?

Pupil: Electricity.

T: What do you mean "electricity"?

P: An electric current.

T: Good – an electric current flows round the circuit. What pushes the current?

P: The battery. It's got a voltage.

T: Ah! The voltage between the two terminals of the battery pushes the current around the circuit. What happens if we use more than one cell?

P: The bulb is brighter.

T: Why?

P: More volts. (contd. on page 18)

T: There is a bigger voltage between the ends of the two batteries. What makes a lamp glow brighter?

P: A bigger current.

T: So. The bigger the voltage the bigger the current flowing round the circuit. What we want to do is to try to find out if the size of the voltage and the size of the current are connected in some way if we use different voltages.

P: We can't tell how big the current is with a lamp. It needs an ammeter.

T: Good. We need instruments to measure the voltage and the current. What else do we need?

P: A lot of batteries and some wire.

T: We haven't got a lot of batteries – at least not enough for everyone.

P: Use labpacks.

T: Exactly. Now, what about the piece of wire? If the experiment is going to be fair what things will we need to control? That means what things should we keep the same during the experiment?

P: The piece of wire.

T: What things about the piece of wire do we need to keep the same?

P: Use the same piece of wire.

T: We can still change things even with the same piece of wire. What about its length? . . .

The discussion continues.

It is likely that many teachers, especially those who have been involved in the Nuffield schemes, will be familiar with this approach. Probably even more will have used it with younger pupils as it figures in a number of published schemes for lower school science. The important thing is that we continue to use the technique as a matter of course for all practical work. It is likely to pay dividends in a number of ways. As well as developing design skills, pupils should become more involved in experimental work and be more aware of what they are doing and why they are doing it. This can lead to an enhancement of performance in all aspects of practical work.

Summary

We have described a number of alternative management strategies which provide the teacher with a range of possibilities for coursework assessment. The nature of some skill areas will dictate special organisation and cause more disruption. The teacher must plan the assessment programme to minimise such occasions by making full use of those strategies which facilitate simple and rapid assessment.

As we have seen, however, not all strategies are suitable for the assessment of every skill area. Table 3.4 summarises the skill areas and the most suitable strategies.

Table 3.4   A skills and appropriate strategy matrix

	Written tests	"Normal" practical work	Formal tests	Demonstrations	Stations	"Novel ways"
Manipulation/following instructions		✓	✓		(1)(2) ✓	
Observation and measurement	(1) ✓	✓	✓	✓	✓	
Recording data	(1) ✓	✓	✓	✓	(2) ✓	✓
Reporting		✓	✓	✓		(1) ✓
Processing and analysing data	✓	✓	✓	✓	(2) ✓	✓
Design and carry out an experiment		(3) ✓	✓			

Notes:
(1) Not wholly suitable.
(2) Restricted to lower performance levels.
(3) Accommodation problems likely.

4 Designing assessment exercises

Introduction

Now that we have an understanding of what needs to be assessed, in terms of the skills and the performance descriptions, and a knowledge of the management strategies which are available, we are set to begin the development of a scheme of assessment exercises.

The first stage is outline planning of the programme. In GCSE we are concerned with coursework assessment rather than practical testing, so it is important that the assessment exercises should fit in with the other elements of coursework as far as possible. The syllabus must be rearranged into a teaching order, and then a teaching scheme produced. The scheme will show: the material to be covered, the methods to be employed, and the experimental work to be carried out.

Many teachers will be familiar with the production of such teaching schemes but for GCSE a new consideration is desirable. With the emphasis on processes, the scheme should include references to the syllabus assessment objectives. This is particularly important in the case of practical skills as the teacher must be satisfied that adequate opportunity has been allowed for the acquisition and development of skills before they are assessed. Table 4.1 illustrates part of a teaching scheme for the NEA syllabus A which embodies these ideas.

A complete teaching scheme set out in the manner of Table 4.1 will allow the teacher to map out an assessment programme. This procedure has the advantage over packaged assessment schemes in that it takes the particular circumstances of the centre (such as apparatus available, group size and laboratory accommodation) into account from the outset.

Once the practical work has been outlined, a more detailed analysis of each practical exercise in terms of the apparatus and the skills involved can be made. This can be integrated into the teaching scheme to produce a detailed "map".

The map will show when skills are introduced, when opportunities for development exist and when opportunities for assessment within normal coursework might exist. I say "might", because there are further considerations before we embark on assessment within coursework.

Table 4.1   A sample teaching scheme including practical skills

Syllabus topic	Teaching method	Objectives	Relevant apparatus and skills in practical work	Teaching materials and additional information
3.1 Forms of Energy	1 Teacher demonstration – see worksheet 2 Accounts and discussion of observations 3 Written exercise on recognising energy forms	Observation Knowledge/recalls Reporting	Observation, safety	Reading *Nuffield Book 3* Worksheet
3.2 Energy Changes	1 Practical circus 2 Account of observations 3 Demonstration/discussion: steam engine	Observation Recall Reporting	Follow instructions Observation, reporting Safety	*Reading about Science Book I* Science 2000: *Unit 3*
3.3 Conservation of Energy	1 Discussion in terms of energy circus	Understanding Interpretation	—	
3.4 Measuring Energy Transfer by Work Done **3.5 Use $W = f \times d$** **3.6 Joule as a Unit**	1 Introductory demonstration and discussion 2 Class practical using $W = f \times d$ 3 Class teaching and discussion problems 4 Class practical: Transferring energy using a pulley system	Application Knowledge Application	Following instructions Measurement: metre rule, newtonmeter Presentation of data Presentation and drawing conclusions	*Physics +*: Hay Inclined Plane, Physics and Sport *Simply Physics Nat Phil "O"*, Ch. 9
3.7 Use $KE = \frac{1}{2} mv^2$ **3.8 Use $\Delta PE = W \times \Delta h$**	Class experiment: Converting PE to KE	Recognising misconceptions	Measuring mass and distance Using ticker tape timers Following instructions	(Level Q candidates only)
3.9 Power in Watts **3.10 Use Equation:** **Power = $\dfrac{\text{Energy}}{\text{Time}}$**	1 Class teaching 2 Class work sheet 3 Class demo power output of pupil Class expt. measuring power output of electric motors and toy cars	Knowledge Understanding Application	Instructions Use of stopwatch Measuring mass, distance, use of stopwatch	*Reading about Physics*: Physics and Sport, The Bicycle See *Nuffield*, Year 4

Although individual skill areas might be involved in a particular exercise, it may not be possible to make an assessment for a number of reasons. The class organisation might not lend itself to making assessments. A less obvious point, but one which is very important, is that the content of the exercise may mean that it is not possible to make assessments at all of the levels of performance which are specified by the syllabus. For example, many schemes differentiate performance in reading scales by the complexity of the scale. Exercises which do not involve an adequate variety of scales would not allow assessment at all levels.

Once we have an overall view of the assessment opportunities within normal coursework we can begin to look at the need for special activities and alternative methods of class management as a means of completing the programme.

Differentiation and discrimination

One of the key concepts in GCSE assessment is "positive achievement". In the sciences, the opportunities for candidates to demonstrate "what they know, understand and can do" are to be provided by "differentiated" papers or questions.

Differentiation means that the exercises provided will be at different levels of difficulty according to the ability of the candidates. The intention is that candidates should be able to show evidence of achievement and thus they should score relatively highly if the level of the task is appropriate. Indeed, the SEC guide for the sciences states "there might well be doubts about a scheme on which candidates could receive Grade F without having demonstrated success on more than half the tasks presented to them". In other words, if we have set appropriate tasks, candidates of average ability (those who will achieve a Grade F) should score at least half marks.

Many teachers will find it difficult to shift from the traditional idea of discrimination – the idea that assessment should produce a wide spread of marks – to differentiation, but as the idea of positive achievement becomes familiar, we should find the situation where almost all of the candidates have scored above half marks more acceptable.

Differentiated exercises involve a measure of prejudgement on the part of the teacher. It might be argued that coursework assessment allows for further assessments to be made if the teacher has judged wrongly, but the practical problems involved in assessment will limit the opportunities.

There is also an additional problem. For those assessments which require direct observation, it is not possible to assess all candidates in a particular group on the same occasion and, it follows, using the identical exercise. The teacher is then faced with the problem of ensuring that exercises are of equal difficulty. In practice, this should not be too difficult as long as the factors which make tasks difficult are borne in mind.

Guidance on what makes tasks difficult is available from APU reports. Further help can be obtained from a detailed study of the grade and performance descriptions provided by the examining groups.

A study of all of the schemes can provide the teacher with a summary of the views of the large number of teachers who have been involved in the production of the syllabi.

Achieving differentiation

There are three ways in which we might go about achieving differentiation. The most obvious way is to design exercises which are aimed at specific achievement levels. In some syllabi the performance descriptions are framed in terms of the difficulty of the task and so there is no alternative. Such syllabi are a useful guide, even for centres who are not following them.

Ideally, teaching groups will have a limited range of abilities and so common exercises could be used. However, in groups in which there is a wide spread of ability, we will be faced with the problem of using different exercises with different members of the group.

It is undesirable to attempt to carry out such assessments on the same occasion because of the difficulty of matching performance on different tasks to the criteria. The teacher would be faced with the problem of remembering two sets of criteria and switching between them, and this in addition to all of the usual problems which beset practical work! A more satisfactory approach would be to use the lower level task for assessment in one session. One could then use a higher level task in a similar area of content for those suited to it on a later occasion.

The criteria which involve help as a discriminator might well be assessed using common exercises since it could be argued that the giving of help changes the level of the task.

A second way of achieving differentiation is by outcome, using what might be termed "neutral stimulus" exercises. In these, a common problem is set but candidates produce different levels of solution according to their abilities. Experimental design is an obvious area where this approach is applicable.

The third way is to use stepped exercises in which there is an incline of difficulty in the stages of the exercise. Many physics practical sessions consist of a series of operations, each of which might not be difficult on its own, but when put together to form a complete experiment might well cause less able candidates problems.

The sequential nature of much of the experimental work which is traditionally carried out, means that many exercises cannot be stepped with operations in order of increasing difficulty. It is not possible to rearrange the sequence. If difficult operations are involved near the beginning of the exercise, there must be some way of enabling less able pupils to get past these stages in order that they may be able to show evidence of success in the skill areas involved in the later stages.

The adoption of schemes which involve separate skill areas tends to preclude stepped exercises unless they are conceived as a sequence of completely separate exercises which become progressively more difficult. Such an approach would not be appropriate for stations because all candidates must start at the lowest level and follow the same sequence. Instead, every candidate would be provided with the same series of tasks at the same station. The most obvious area for this approach is the assessment of measuring skills and sample exercises 10 and 11 illustrate this.

The exercises allow an assessment at all skill levels in the LEAG syllabus B, but a suitable choice of the instruments involved, to allow the necessary range of types of scale, would allow the exercises to be used to assess this skill area for the other schemes.

Isolating skill areas

It is important in an extended exercise which is made up of a number of stages, that performance in the parts which are to be assessed is not dependent on a successful outcome in the earlier stages.

For example, it would be undesirable to attempt an assessment of the ability to record or process data in an experiment where the candidate fails to obtain any data! For candidates whose performance in some of the skill areas is poor, it is important that steps are taken to ensure that such a situation does not arise.

One way of ensuring this is to isolate the skill areas, wherever possible, so that the assessment exercise consists of the performance of a restricted number of skills or processes. The stations strategy is an obvious method of achieving such an end.

Even with able candidates it is likely that the use of exercises which isolate the skills in this way will be advantageous as far as making the assessment goes. The removal of the need to determine the status of performance before the candidate commences a part of the exercise which is to be used for assessment, will make the assessment easier. It should also increase the number of assesments which can be made in a session by reducing the time spent by the teacher in checking each candidate's status.

In situations where the skills are not isolated, it is important that a check be made before each assessment so that the teacher can be certain that assessment will be valid.

Another important point to remember is that if the assessment involves direct observation of one part of an experiment, it is likely that many of the candidates will reach this point at about the same time. Holding up the practical work so that the teacher could make a large number of assessments would be a bad thing and should be avoided. However, the number of assessments which could be made could be increased by assessing different candidates on different skill areas at different stages in the experiment.

Reducing factual and content load

The APU reports suggest that when

"... pupils are required to apply science concepts which they have been taught in other science contexts performance is low."

and although this was in reference to pencil and paper tests, it is likely that similar results would be obtained in practical work.

An attempt to assess performance of a process or skill would not be valid if the candidate required additional knowledge and a number of other skills in order to carry out the exercise successfully. A further problem arises when the instructions contain references to concepts which candidates find difficult. An understanding of the concepts might not be necessary, but less able candidates might be unable to extract the relevant parts of the instructions. On the other hand, it would be wrong to attempt to differentiate exercises by building them around more difficult concepts, rather than more difficult aspects of the skills.

An example of an exercise which treats a traditional subject, that of density, in a way which avoids the concept is shown in sample exercise 1. This approach does not argue for the elimination of "difficult" concepts from physics at this level. It does, however, mean that we must reconsider the idea that one of the main aims of practical work is to support and develop an understanding of theory.

Language

Physics texts and examination questions are notorious for the high language level of understanding which they demand. The problem of language is twofold. The special vocabulary and the precise way in which physicists use words which have different meanings in everyday usage is one problem. A knowledge of such vocabulary is one of the objectives of all of the syllabi and one would not expect it to be eliminated from instructions for practical work.

On the other hand, much written material uses difficult language even when it is not using specifically scientific words. We must be sure that the language level of instructions is simple enough to enable all candidates to understand what they have to do. This is especially important when we are assessing skill areas which involve the ability to follow instructions.

As a general guide, the following points should be kept in mind when producing worksheets:

- The reading level should be two years below the reading age of the candidates. Keeping sentence length short should help to achieve this.
- The material should be clearly set out with each instruction on a new line.
- The structure of the worksheet should reflect the structure of the exercise.
- Care should be taken to ensure that the syntax is clear and unambiguous.
- Diagrams should be clearly drawn and follow the conventions with which the pupils are familiar.

It will also help if a standard format and wording is adopted wherever possible. This accustoms pupils to working to a routine and helps build confidence. The approach has been adopted in the sample exercises in the next chapter.

A few minutes' review of the draft with colleagues should help to ensure that the material is suitable.

If the teacher is unfamiliar with the problems of producing worksheets, especially for less able pupils, it is advisable to read through the worksheet with the group before they start the exercise. Any problems which come to light at this stage can be used to modify the worksheet for future use. The important thing is to be sure that lack of clarity in the instructions is not the cause of a low level of performance.

There is an additional problem which is not connected with language but which must be considered when writing the worksheets. This is whether or not we tell the pupils which skill areas are being assessed?

In exercises where only a part of the performance is being assessed, this might lead to candidates placing undue emphasis on the relevant parts to the detriment of the rest of the exercise. As a general rule it is probably desirable that candidates should be aware of the skills involved and that assessment may take place, but it is important that the teacher stresses the point that neglect of other parts of the experiment may well result in a poor performance in the assessment.

Mathematical requirements

A number of the skill areas involve what might be termed mathematical skills. The presentation of data and the handling and analysis of data are obvious examples.

All syllabi include a statement of the mathematical abilities expected of candidates at the different grade levels. It is important that these are borne in mind when designing assessment exercises so that the mathematical demands are appropriate to each candidate's abilities.

There are however some problems which might be introduced if careful attention is not paid to the relationship between the mathematical requirements and the performance descriptions provided by some schemes.

As an example we will consider the LEAG syllabus B. The mathematical requirements are differentiated (in common with most other schemes) in such a way that some abilities which are important in experimental work, especially in the area of what is termed "graphicacy", are not expected from grade F and G candidates. Probably the most significant ability is the plotting of points which do not lie on a crossover of key scale markings, the most probable situation in experimental work. Again, drawing "the best fit line where points fit less perfectly" is a situation which grade F and G candidates are likely to find is the norm using the results of practical work but which they are not required to be able to do in the written paper.

The other most important point is probably the provision of predrawn axes and scales for graph axes. For those schemes which mention such provision in the performance criteria, the teacher should prepare suitable blanks before the assessment commences.

In addition, the requirements for written examinations, such as the availability of calculators and the provision of the relevant formulae in an appropriate form should be adhered to in coursework assessments.

Marking and recording assessments

The need to make adequate records of the assessment must be built into the exercise as it is designed. We have already mentioned the need for endchecks to reduce the amount of direct observation required. It is equally important that a suitable method of recording the assessment is built into the other areas from the outset.

There are two ways of organising the marking. In exercises which produce a written record the marking can be carried out after the exercise is completed. Alternatively, in skill areas where the teacher is making a direct observation of performance, some form of checklist or grid will provide the simplest and most effective way of recording. It should be remembered that discussions with the pupil can provide additional information for the assessment, and for many pupils this may well be the most effective way of obtaining such information.

In either case, we are faced with the problem of fitting the marks to the performance descriptions provided by the syllabus. There are two approaches to this problem.

The performance might be marked to a mark scheme, in which a mark is awarded for each point and then the total is scaled to the syllabus mark scheme. This approach might be appropriate where the syllabus has extended descriptions of performance levels, or which includes a relatively large number of discriminators. It would be impossible to design exercises which cover all aspects of performance or involve all of the discriminators.

Table 4.2   A matrix for the assessment of performance at Handling Experimental Data on the MEG scheme

Candidate name	Recording data	Graph plotting	Conclusions	Total	Scaled mark
	/5	/5	/5	/15	/10
BRADLEY	3	4	2	9	6
CLARK	4	4	3	11	7
COOK	5	5	5	15	10
COX	2	3	1	6	4
DAVIS	5	5	4	14	10
HILL	4	3	3	10	7

An example is shown in Table 4.2. This is based on the MEG skill area of Handling Experimental Data. Each performance description involves three areas and there are five levels for each area. Candidates are marked out of five on each area using the performance descriptions provided in the syllabus and marks are then aggregated and scaled.

The important factor is that the points on the mark scheme should be interpreted in terms of the statements included in the performance descriptions, and there must be a correlation between the hierarchy in the mark scheme and that in the performance descriptions.

In schemes where simple statements of performance are provided, such as in the NEA and SEG schemes, there is no alternative but to use the performance statements as the mark scheme. What is important is that the exercise is designed to enable all levels of performance to be demonstrated. Thus it would not be acceptable to attempt to assess a candidate's ability to select apparatus using only two alternative pieces of apparatus when the scheme allows for three marks.

With the relatively simple range of performance descriptions involved in these schemes, a simple matrix of skill areas against candidate names will be adequate, especially as teachers become more familiar with the scheme.

A checklist or matrix is a useful way to record marks because it provides a permanent record with immediate access to information as regards what has been done and what remains to be assessed for each candidate. Table 4.3 shows a matrix for the NEA scheme which could be used to assess sample exercise 8.

The teacher managed to observe Briggs and Layton working individually and assessed their performance at following instructions. Hall and O'Donnell were each observed making measurements and a number of candidates' records of results were assessed.

This example highlights the fact that even using the

same exercise, it is not necessary to assess each candidate on the same skill area.

Table 4.3   A matrix for the assessment of sample exercise 8 on the NEA scheme

	Following instructions	Using scales	Selecting apparatus	Recording data	Design experiment	Conclusions
BARBER	–	–		3		
BRIGGS	2	–				
CLARK	–	–	NOT APPLICABLE	3	NOT APPLICABLE	NOT APPLICABLE
HALL	–	3		3		
CLAYTON	3	–		–		
O'DONNELL	–	2		–		
POPE	–	–		3		
ROSS	–	–		2		

An alternative approach is to break down each skill into a number of areas and then use a matrix as a means of recording performance in each area. A sample matrix is shown in Table 4.4 which has been designed to assess sample exercise 4 on the LEAG syllabus B area of use and analysis of data.

Table 4.4   A matrix with skill breakdown for sample exercise 4

	Results different: recognises acceleration	Plots graphs	Constant acceleration	Compares accelerations by calculating slopes	Highest mark
ATKINS	√	√	√	–	3
BARNABY	√	√	√	√	4
COOPER	√	√	–	–	2
HANCOCK	√	–	–	–	1
O'BRIEN	√	√	√	–	3
PATEL	√	√	√	–	3

The exercise only allows assessment up to skill level 4 since the relationship involved is not sufficiently complex for skill level 5.

Once the teacher has a full record of the candidate's work on the checklist the exercise can be assessed in terms of the performance descriptions provided by the syllabus. In most cases this should be a direct transfer if the exercise and assessment scheme are properly designed at the outset.

If there is any difficulty in the process then it is likely that the exercise needs redesigning.

Moderation

The purpose of moderation is to ensure consistency of standards within centres and between centres. The most important point, as far as the centre is concerned, is to try to ensure that the standards of exercises and assessment meet the performance descriptions set out in the syllabus and that the necessary records are kept to support the centre's assessment.

Even when there is no requirement stated in the syllabus, it is important that all of the records of assessment are retained for possible scrutiny during the moderating procedure. Proper records will support the centre's case if there is any dispute about moderation. In addition it is important that records are available when pupils change schools or teachers so that the new teacher has as much information as possible. Even if the candidate is to study under a different syllabus, comprehensive information of experiences and abilities could save the new teacher (and the pupil) a great deal of time!

The training provided by the examination groups should have provided teachers with details of the moderating procedure. If there are any unclear points then it is essential that centres clarify them by contact with the examination group.

Summary

We have considered the points to be borne in mind when designing exercises for assessment. The stages in the development of an assessment programme are outlined in the flowchart, Fig 4.1 on page 24.

It may seem an impossible task when considered on a theoretical basis, but a few concrete examples and discussions of the product with colleagues, should soon provide teachers with the experience and confidence to produce the sorts of exercise which will make the fears of assessment a distant memory!

Fig. 4.1   The processes involved in the production of an assessment scheme for GCSE Physics

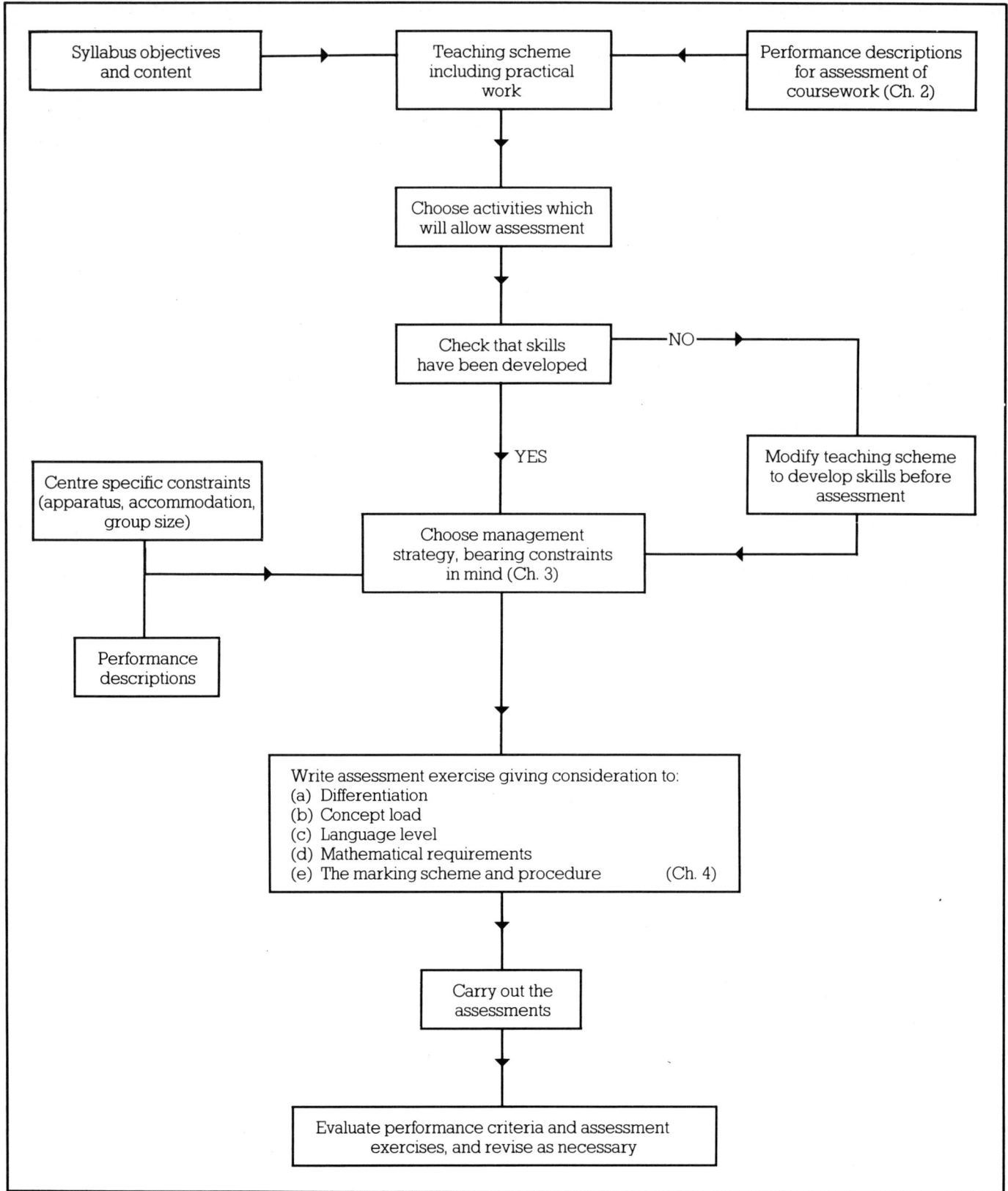

5 Specimen assessments and teaching notes

This chapter includes a number of specimen exercises which can be used for assessments for most of the syllabi. The exercises have been designed to use some of the strategies which we have discussed. For some of the strategies, however, sample worksheets are inappropriate. There are no exercises designed specifically to be used as demonstrations, although some of them could be if desired. Teachers should be able to produce their own exercises using the guidelines provided in the previous chapter.

Wherever possible, the exercises use standard laboratory apparatus. Special apparatus and materials are kept as simple as possible to minimise the amount of extra work involved.

The exercise can be modified to suit the apparatus available in the centre, or to suit the requirements of the group as perceived by the teacher. No one knows better than the class teacher what is most appropriate for a particular group.

The exercises have been kept relatively straightforward to ensure that they are simple and reliable to operate. The pressures of practical assessment are enough without the teacher having to become familiar with a completely new range of experiments. A number of exercises, however, do have a novel appearance and these could be adapted to provide additional exercises for assessing experimental design.

The exercises can be used by centres following schemes which allow the assessment of individual skill areas, and teachers can select the skill areas to be assessed. In addition, most of the exercises can be used by centres following the SEG syllabus. The introduction to the worksheet includes a brief note on the background to the experiment and this, together with a statement of the problem, could be given to candidates. The worksheet would be available for candidates who are unable to design a satisfactory experiment.

It is not necessary to assess all of the skill areas involved in each exercise. The teacher can select those areas which are most appropriate.

The times given can only be rough estimates. The actual time will depend on the ability of the candidates and their experience as well as the amount of prior preparation and technical help available.

A number of the exercises are designed to provide opportunities for experimental design. In a number of these there are alternative problems depending on the variables to be investigated, and provision is made in the pupil worksheets for the teacher to adapt them. The majority of these exercises are also provided with additional help sheets which can be given to candidates to assist them if necessary.

For the experimental design exercises, the time will depend very much on the complexity of the strategy adopted and on the persistence of the candidates. The organisational problems and the time taken to get the experiments underway will, however, be reduced if the most likely apparatus is readily available. Suggestions of the apparatus which might be needed are given in the teacher's notes.

It is not possible to provide mark schemes for each exercise because of the wide variation in the performance criteria which have been adopted by each scheme. A number of marking points which can be used in the construction of a marking scheme are provided for each exercise.

The method of marking will depend on the skill areas being assessed and the strategy adopted as well as the particular syllabus being followed. If the teacher is using direct observation to assess some skill areas then a matrix constructed following the guidelines set out in the previous chapter will be required.

Each exercise is accompanied by teacher's notes which provide:

- the skill areas which might be assessed
- the previous experience, skills and knowledge expected of the candidate
- the management strategies which might be adopted
- an estimate of the time required by the candidate
- the resources required for the exercise and details of advance preparation
- suggested modifications or adaptations to the exercise, where appropriate.

Teacher's Notes

Exercise 1

Skills tested

Following instructions/manipulation
Measurement skills – either by direct observation or by reference to pupil records if the teacher has a record of the masses and volumes of the objects.
Selection of equipment
Presentation of data
Drawing conclusions
Reporting

Previous experience

This exercise involves measurement of mass and volume for a number of solid objects. It does not require prior work on density and so could be used at any time after pupils have acquired the necessary manipulative and measuring skills.

Candidates of all levels of ability should be capable of demonstrating positive achievement in this exercise.

The exercise requires:
Measurement of volume of regular and irregular solids using measuring cylinders, displacement can and rule.
Use of a balance.

Management strategy

The exercise can be used either for individual work in normal practical work, as a formal test or as a stations exercise with each sample at a different station. If different sets of materials are used, stations could alternate as a means of reducing the chance of pupils swapping information.

Time required

This will depend on the strategy adopted and the number of balances available. With an adequate number of balances or using an electronic balance (which would exclude making a full assessment of measuring skills), pupils should be able to complete the exercise in a maximum of 40 minutes.

Resources

Each candidate will need 5 samples of a material which is denser than water. If possible, one sample should have a regular shape so that its volume can be determined from measurements of its dimensions.

The samples should be of a suitable size to fit into the measuring cylinders and/or the displacement can provided. They should be identified by a number painted on them. If every object has a different number, the pupil records should allow assessment of measurement skills.

Suitable materials are:

glass rod, tubing, stoppers, blocks
metals (aluminium, steel or copper) – these can be scrounged from the craft department or pieces collected from around the house – copper or aluminium pipe, coins, nails etc.

In addition each candidate will need:

A range of measuring cylinders
A displacement can
A beaker for water
A rule
Lengths of thread and string
Access to a balance
Graph paper

Spare measuring cylinders (preferably plastic) should be available as some objects might get stuck and time is not free for the teacher to dislodge these.

Marking the assessment

Some or all of the following marking points can be used to develop marking schemes according to the scheme used and the areas assessed.

(a) Care in measuring volumes
 – avoiding spillage
 – avoiding air bubbles
 – using thin thread to suspend objects
 Selection of optimum method of measuring volume and optimum measuring cylinder
 Selection of most sensitive scale if dual-scale balance is used
 Measurements correct to within the sensitivity of the best instrument available

(b) Table drawn with
 – columns labelled
 – units included
 – all measurements included
 – all data easily accessible
 – derived quantities correctly calculated

(c) Graph drawn with
 – correct axes labelled including units
 – suitable scale
 – points correctly plotted
 – best straight line drawn

(d) Statement of connection between mass and volume or calculation of ratio of mass to volume

Exercise 1. The Connection between Mass and Volume

Before you begin:

Read through all of the instructions carefully.
Make sure that you understand what you have to do.

You are going to measure the mass and the volume of each of 5 numbered objects. They are all made from the same substance.

Make each of your measurements in the most accurate way that the apparatus provided will allow.

Record all of your measurements in a suitable table.

Apparatus   You are provided with:

A set of 5 numbered samples
A range of measuring cylinders
A displacement can
A beaker for water
A balance (you may have to share a balance)
Lengths of thread and string
Rule, graph paper
Check that you have all the apparatus listed.

Method

1. Measure the mass of each of the numbered objects. Record your results in a table.

2. Measure the volume of each of the objects and include these measurements in the table.

There are different ways of measuring the volume of an object. Choose the one which you think is best for each object.
One method of measuring volume is given below.

3. If you use a measuring cylinder, write down the maximum volume it will measure in the table.

4. Check that you have measured the mass and the volume of all 5 objects.

5. Plot a graph of the mass of each object against its volume on the graph paper provided.

6. Use your graph to decide if there is any connection between the mass and the volume of the objects.

7. Write a report describing exactly what you did.

Measuring the volume of an object by displacement

(a) Put a quantity of water in one of measuring cylinders.
    Make a note of the volume.

(b) Tie a piece of thread or string around the object.
    CAREFULLY lower the object into the measuring cylinder until it is completely covered by the water.
    Make sure that you do not splash the water and that there are no air bubbles.

(c) Measure the volume of the object plus water.

(d) The volume of the object is the difference between the two measurements.

Teacher's Notes *Exercise 2*

Skills tested

Following instructions/manipulation
Presentation of data
Drawing conclusions
Reporting

Previous experience

This exercise could follow on from an experiment measuring load and elongation values for a single spring. It is relatively simple and could be used to assess less able pupils.

The exercise requires:
that pupils should be familiar with the load–extension behaviour of a spring or elastic band. They should be able to use a set-square to help in the accurate measurement of length of the springs.

Management strategy

The exercise can be used in normal class practical as an individual exercise since it does not place an undue strain on the provision of apparatus. If springs are in short supply, identical rubber bands could be used. A suitable weight for the particular bands used should be determined by prior experiment.

The exercise could be modified slightly for use as a stations exercise. Each station would be set up with different numbers of bands and pupils would be required to make the necessary measurement. This would restrict the skills tested to presentation of data and drawing conclusions but it would effectively isolate the skills tested.

Time required

As an individual exercise pupils could carry out the practical work in about 20 minutes. If the experiment is set up as a stations exercise, a group of pupils could carry out the set of stations in less than 10 minutes.

Resources

Each pupil will need 4 identical expandable springs or elastic bands. They will also need 3 split rings made from stiff wire to join the springs together. A quick way to produce the rings is to coil stiff wire around a clamp stand and then make a saw cut along the length of the coil.

The only advance preparation needed is the making of the rings and checking the weight needed to produce a reasonable extension of the rubber bands used.

In addition, each pupil will need:

A 100 g mass with a means of hanging it from the springs
A metre rule
A clamp stand, boss and clamp
A set square
Graph paper

Alternatives or extensions

A similar exercise could involve springs suspended in parallel using zig-zag wire to prevent the springs from slipping side ways. This produces a more complex relationship between number and extension and would be suitable for the assessment of the ability of more able candidates to draw conclusions.

Marking the assessment

Some or all of the following marking points can be used to develop marking schemes according to the scheme used and the areas assessed.

(a) Assembly and use of apparatus (by observation)
   - stand and clamp secure
   - rule clamped vertically
   - spring suspended in suitable place
   - care in measuring end of spring

(b) Table drawn with
   - columns labelled
   - units included
   - all measurements included
   - all data easily accessible
   - extensions correctly calculated

(c) Graph drawn with
   - correct axes labelled including units
   - suitable scale
   - points correctly plotted
   - best straight line drawn

(d) Statement of connection between extension and number of springs

Exercise 2. The Extension of Springs in Series

Before you begin:

Read through all of the instructions carefully.
Make sure that you understand what you have to do.

When you hang a weight on a spring the length of the spring increases. The **increase** in the length of the spring is called the EXTENSION. You have already found out that, for a single spring, the extension depends on the size of the weight.

In this experiment you are going to try to find out how the extension is affected by using different numbers of springs connected in line.

Make each of your measurements in the most accurate way that the apparatus provided will allow.

Record all of your measurements in a suitable table.

Apparatus   You are provided with:

4 springs
A 100 g mass (a weight of 1 N)
A metre rule
3 rings for connecting the springs
A clamp stand, boss and clamp
A set square

Check that you have all of the apparatus listed.

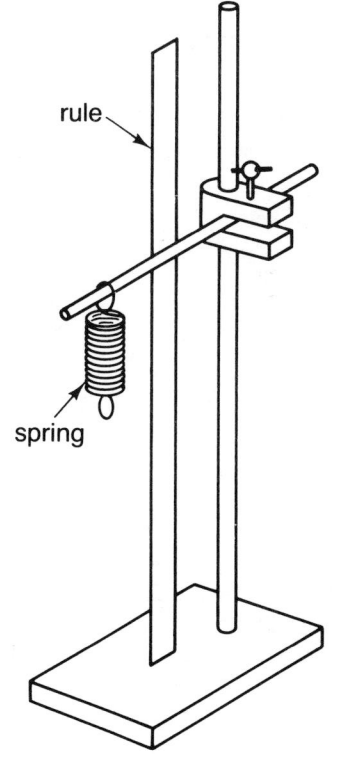

Method

 1. Set up the apparatus as shown in the diagram.

 2. Measure the position of the bottom of the spring as carefully as you can and make a note of it.

 3. Hang the weight from the spring. Measure the new position of the bottom of the spring. Make a note of this.

 4. Work out the extension of the spring.

 5. Join another spring in line using one of the split rings.

 6. Measure the position of the bottom of this spring and make a note of it.

 7. Hang the weight from the lower spring. Measure the new position of the bottom of this spring. Make a note of this.

 8. Work out the extension of the springs.

 9. Repeat steps 5 to 8 for 3 springs in line and 4 springs in line.

10. Plot a graph of extension against the number of springs.

11. Use your graph to decide if there is any connection between the extension and the number of springs.

12. Write a report describing exactly what you did.

Teacher's Notes *Exercise 3*

Skills tested

Following instructions/manipulation
Presentation of data
Drawing conclusions
Reporting

Previous experience

This exercise is a variant on the more usual "Hooke's Law" type of experiment. It is straightforward and does not require any prior knowledge about elasticity or beams. It could be used as an alternative to exercise 2 in the same session.

The exercise requires:
Measurement of length

Management strategy

The exercise can be used for individual work in normal practical work since it does not require much equipment – just a ready supply of masses and rules.

Time required

About 20 minutes for the practical work.

Resources

Each pupil will require 2 blocks of rigid material which have the same dimensions. Blocks from the Nuffield materials kits or even housebricks would be suitable. Other strips of wood could replace the metre rule for the beam if necessary. CHECK that they will deflect a measurable distance under a reasonable load.

In addition, each pupil will need:

A half metre (or metre) rule
6 masses (either a range of masses between 0.1 kg and 1 kg, or 6 × 100 g stackable masses
Graph paper

Alternatives and extensions

The exercise could be altered to investigate the effect of span on deflection for a given load. A different exercise could involve the rule clamped at one end to produce a cantilever.

Problem-solving exercises could be developed from the basic exercise. An investigation of the effects of width or depth of beam on deflection, or the number of similar beams (strips of spaghetti?), are possibilities.

The alkathene strips from electrostatics kits or thin strips of balsa wood show an appreciable depression. These would be shorter than a metre rule and so would be easier to manage for large classes.

Marking the assessment

Some or all of the following marking points can be used to develop marking schemes according to the scheme used and the areas assessed.

(a) Assembly and use of apparatus (by observation)
  – stable assembly
  – correct positioning of beam
  – correct position for measurements
  – care with loading beam at centre

(b) Table drawn with
  – columns labelled
  – units included
  – all measurements included
  – all data easily accessible
  – derived quantities correctly calculated

(c) Graph drawn with
  – correct axes labelled including units
  – suitable scale
  – points correctly plotted
  – best straight line drawn

(d) Statement of proportionality between load and deflection, or calculation of proportionality constant.

Exercise 3. Bending Beams

Before you begin:

Read through all of the instructions carefully.
Make sure that you understand what you have to do.

When a force is applied to a beam (a long piece of material such as a piece of wood or a girder) the beam bends.

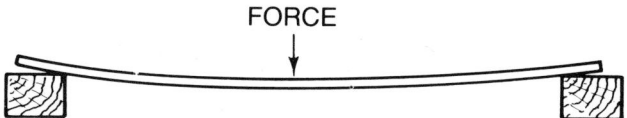

FORCE

You are going to find out the amount of bending produced by different forces.

Make each of your measurements in the most accurate way that the apparatus provided will allow.

Record all of your measurements in a suitable table.

Apparatus   You are provided with:

A metre rule (the beam)
Another rule for measuring
Two supports for the beam
A set of masses for loading the beam
Graph paper

Check that you have all of the apparatus listed.

Method

1. Put the two supports on the bench so that there is exactly 80 cm between them.

2. Put the metre rule across the two blocks as shown in the diagram.

3. Measure the distance from the bench to the top face of the rule at the 50 cm mark.

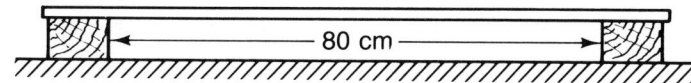

80 cm

4. Put a mass on the centre of the rule to provide a force on it. Make a note of the mass in the table.

5. Measure the new height of the top face of the rule. Make a note of it in the table.
    The difference between the original height (without a force) and the new height is called the DEFLECTION of the beam.

6. Repeat the experiment with 5 different masses on the centre of the beam, giving a total of 6 sets of readings.

7. Plot a graph of deflection against mass using your results.

8. Use your graph to decide if there is any connection between the deflection and the mass.

9. Write a report describing exactly what you did.

Teacher's Notes

Skills tested

Following instructions/manipulation
Presentation of data
Drawing conclusions
Reporting

Previous experience

This exercise is relatively straightforward, but it does provide the opportunity for candidates to demonstrate performance of some of the techniques of measurement such as repeating readings.

The levels of conclusion which are possible range from simple recognition of uniform acceleration to an understanding of the effect of slope on acceleration and calculation of accelerations for the most able pupils. More able candidates should also be aware of the errors involved in making measurements of small time intervals.

The exercise requires:
that candidates should be familiar with the use of ruler and stopwatch. If conclusions are to be assessed they should also have covered work on velocity, acceleration and velocity–time graphs.

Management strategy

The exercise can be used for individual work to assess part of a group. The major problem is the amount of space required by each candidate and the supply of boards.

Stations could be used with different distances marked on the board at each station. This would restrict the range of skills which could be assessed.

Time required

About 30 minutes for the practical work.

During the time in which one group of candidates is processing data a second group of candidates could use the apparatus to make measurements. This would increase the number of candidates who could be assessed.

If a second group is assessed then the time taken should be between 45 minutes and 1 hour.

Resources

Each candidate will require a rigid board or plank with a smooth surface. It should be long enough to allow the object to roll at least 1.5 m.

Candidates will also need a number of equal sized blocks or books, each about 5 cm thick: 4 should be sufficient but the number will depend on the length of the board and the thickness of the blocks. An end stop on the board will prevent the object from rolling on to the floor.

In addition each pupil will need:

A ball bearing, marble or short length of round rod or bar
A metre rule
A stopwatch
Graph paper

Alternatives and extensions

Long lengths of plastic rain gutter or wide bore glass tube could replace the board.

Investigations of the performance of different makes of model car, or the effect of load on the acceleration of a trolley, could be designed around the same method.

Marking the assessment

Some or all of the following marking points can be used to develop marking schemes according to the scheme used and the areas assessed.

(a) Assembly and use of apparatus (by observation)
   - correct slope
   - measurement of height at correct point
   - correct sequence of operations
   - stopwatch started as object is released

(b) Table drawn with
   - columns labelled
   - units included
   - all measurements included
   - all data easily accessible
   - average velocities correctly calculated

(c) Graph drawn with
   - correct axes labelled including units
   - suitable scale
   - points correctly plotted
   - best straight line drawn

(d) Statement of uniform acceleration
   Recognition that acceleration is greater with a steeper slope
   Calculation of accelerations from slope of graph

Exercise 4. Rolling down a Slope

Before you begin:

Read through all of the instructions carefully.
Make sure that you understand what you have to do.

When an object rolls down a slope it accelerates. You can find the average velocity of the object by measuring the time taken to cover a measured distance. You are going to measure the times taken for an object to roll different distances down the same slope. You will then repeat the experiment using a different slope.

Make each of your measurements in the most accurate way that the apparatus provided will allow.

Record all of your measurements in a suitable table.

Apparatus You are provided with:

A long board   A metre rule
Blocks to provide a support at one end of the board   A stopwatch
An object which will roll freely   Graph paper

You will need a calculator to work out your results.
Check that you have all of the apparatus listed.

Method

1. Set up the slope by putting blocks under one end of the board. Adjust the height until the object rolls freely down the slope without being pushed.

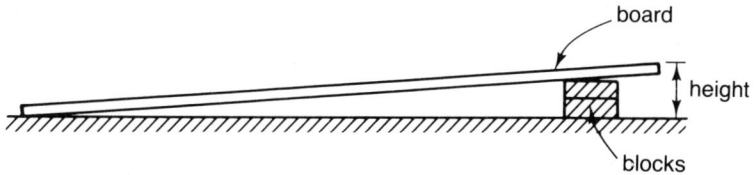

2. Measure the height of the slope as shown in the diagram. Make a note of it.

3. Measure the time taken for the object to roll 50 cm down the slope from rest. Make a note of your measurements in your table of results.

4. Measure the time taken for the object to roll different distances from rest. Distances of 75 cm, 100 cm, 125 cm and 150 cm are suitable. Record your measurements in a table.

5. Use the equation

$$\text{average speed} = \frac{\text{distance travelled}}{\text{time taken}}$$

to work out the average speed of the object over each distance. Record your results in your table.

6. Plot a graph of average speed against time taken.

7. Increase the slope by putting more blocks under the end of it. Adjust the height until it is exactly twice as high as in the first experiment.
Measure the height and make a note of it.

8. Repeat steps 3, 4, 5, and 6 to produce a second set of results for the new slope.

9. What do your results tell you about the motion of the object in each case?

10. Write a report describing exactly what you did.

Teacher's Notes

Skills tested

Following instructions/manipulation
Measurement skills – by direct observation
Presentation of data
Drawing conclusions
Reporting

Previous experience

This exercise provides an opportunity to use a manometer and is also a suitable vehicle to assess the ability of candidates to recognise mistakes involving basic physics and experimental techniques.

The exercise requires:
that candidates should be familiar with the use of a manometer to measure excess pressure. Some work on the behaviour of gases would be an advantage.

Management strategy

The exercise can be used for individual work during normal practical work or as a formal test.

Time required

About 40 minutes for the practical work.

Resources

Each candidate will require a U-tube manometer. This can be constructed from a length of PVC tubing clipped to a board with a half-metre rule as a scale. There should be some means of holding the manometer in a vertical position.

In addition, each pupil will need:

A boiling tube with a bung to fit. The bung should have a single hole with a length of glass tube through it
Rubber tubing to connect the glass tube to the manometer
A beaker large enough to contain the boiling tube immersed in water
A 0–100 °C thermometer
A stirring rod
A Bunsen burner, tripod, gauze and bench protection
Graph paper

Marking the assessment

Some or all of the following marking points can be used to develop marking schemes according to the scheme used and the areas assessed.

(a) Assembly and use of apparatus (by observation)
  – correct assembly of apparatus
  – safe position and stability of apparatus
  – use of bench mat
  – correct use of Bunsen burner to give gentle heating
  – correct use and position of thermometer when making measurements
  – correct procedure when making measurements

(b) Table drawn with
  – columns labelled
  – units included
  – all measurements included
  – all data easily accessible
  – pressures correctly calculated

(c) Graph drawn with
  – correct axes labelled including units
  – suitable scale
  – points correctly plotted
  – best curve drawn

(d) Validity of conclusion based on mis-match or match between student's results and statement
  Reasons given for discrepancy
  Suggestions for improving experiment, which could include: using a narrower tube between manometer and boiling tube, using a narrower tube for the manometer, and using a larger container so that the volume of the heated air relative to the unheated air is larger and the effect of the volume change as the liquid in the manometer moves is minimised.

Exercise 5. Heating Air

Before you begin:

Read through all of the instructions carefully. Make sure that you understand what you have to do.

You are going to use a simple manometer to measure the excess pressure of the air in a boiling tube at different temperatures.

Make each of your measurements in the most accurate way that the apparatus provided will allow.

Record all of your measurements in a suitable table.

Apparatus   You are provided with:

A boiling tube with rubber bung and glass tube

Rubber tubing to connect the glass tube to the manometer

A large beaker

A 0–110 °C thermometer

A U-tube manometer

A Bunsen burner, tripod, gauze and bench protection

A stirring rod

Check that you have all of the apparatus listed.

Method

1. Set up the apparatus shown in the diagram. Take care to ensure that the joints in the tubing are secure and that you can be reasonably sure that they are airtight.

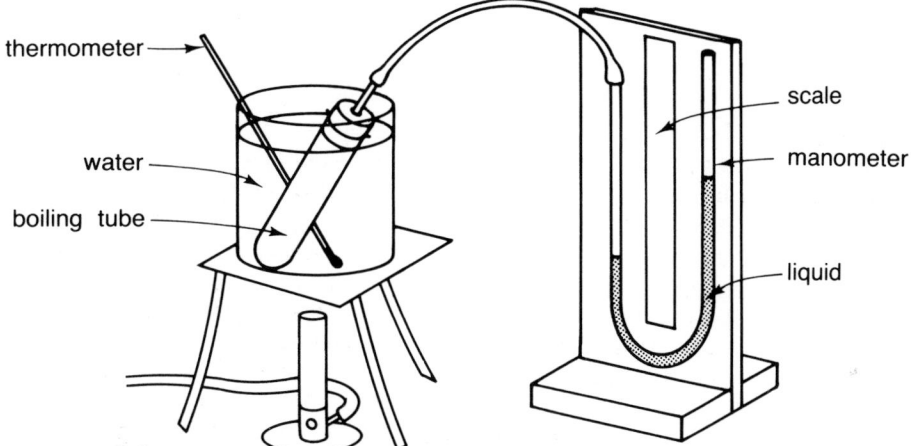

2. Measure the temperature of the water.

3. Measure the extra pressure of the air in the boiling tube.

4. Light the Bunsen burner and heat the water gently.

5. Measure the temperature of the water and the pressure of the air every 10 °C or so.

When you are making a measurement, slide the Bunsen burner from under the beaker and stir the water for about 1 minute so that the air in the boiling tube is at a steady temperature. Be careful to slide the Bunsen burner to a safe distance from where you are looking or writing.

6. Make your last set of measurements when the water is boiling.

7. Plot a graph of pressure against temperature.

8. Write a report describing exactly what you did.

A student claims that the experiment shows that the pressure rises by equal amounts for equal rises in temperature (that pressure rise is proportional to temperature rise).

Do your results support this claim?

Teacher's Notes *Exercise 6*

Skills tested

Following instructions/manipulation
Presentation of data

Previous experience

This exercise is relatively simple. However, less able candidates might have difficulty producing a suitable range of angles of incidence. To make the exercise easier, the teacher could provide a set of duplicated sheets with the position of the prism and the direction of suitable incident rays drawn on them.

The exercise requires:

Measurement of angles. Pupils should be familiar with the technique of producing accurate traces of ray streaks. Candidates should have covered work on refraction, including the terms angle of incidence, normal and incident ray.

Management strategy

Individual or group work. With group work, each pupil could produce a set of sheets and they could then measure the angles and present the data in a suitable format.

Time required

About 25–30 minutes. If duplicated sheets are provided, less time would be required.

Resources

If the teacher decides that pupils should be provided with ready drawn blanks, these should be produced using the prisms to be used by the pupils, and then duplicated.

Each candidate or group will also need:

A raybox with single slit and suitable power supply
A red or green filter
A 60° glass prism
A rule
A protractor
Sheets of plain paper
Graph paper

Each pupil will also need a sharp HB or H pencil.

Alternatives

Different prisms (45°, 45°, 90° or 30°, 60°, 90°) could be used instead of the 60°, 60°, 60° prism.

Marking the assessment

Some or all of the following marking points can be used to develop marking schemes according to the scheme used and the areas assessed.

(a) Assembly and use of apparatus (by observation)
  – ray box correctly connected to power supply
  – correct use of power supply
  – ray box set to produce narrow ray streak
  – care in marking ray positions

(b) Table drawn with
  – columns labelled
  – units included
  – all measurements included
  – all data easily accessible
  – angles of deviation correctly calculated

(c) Graph drawn with
  – correct axes labelled including units
  – suitable scale
  – points correctly plotted
  – best curve drawn

Exercise 6. Bending Light

Before you begin:

Read through all of the instructions carefully.
Make sure that you understand what you have to do.

When a ray of light passes through a prism,
refraction causes it to bend. The ray travels
along a different direction when it leaves the
prism. The angle through which the direction
changes is called the ANGLE OF DEVIATION.

The diagram shows this.

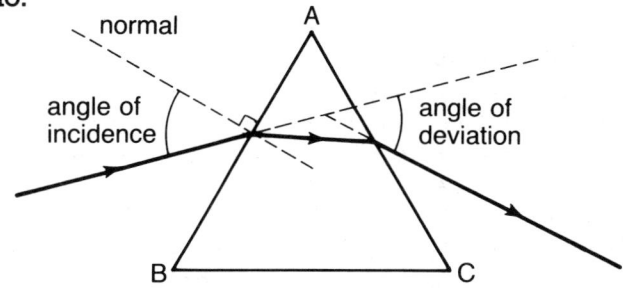

You are going to measure the angle of deviation of a ray of light for different angles of incidence
on the first face of the prism.

Make each of your measurements in the most accurate way that you can.
Record all of your measurements in a suitable table.

Apparatus   You are provided with:

A 60° prism   A coloured filter   A rule   Graph paper
A ray box and power supply A protractor   Sheets of plain paper

You will also need a sharp pencil.
Check that you have all of the apparatus listed.

Method

1. Connect the ray box to the correct power supply. Fit the coloured filter to the front of the ray
   box. Make sure that you can obtain a sharp ray streak.

2. Put the prism on a sheet of plain paper. Carefully draw round it.

3. Place the ray box so that the ray streak hits one face of the prism and leaves the prism
   without being reflected inside the prism. These are the faces marked AB and AC in the
   diagram.

4. Carefully mark the positions of the incident ray and the emergent ray.

5. Remove the ray box and prism and use the rule to draw the incident ray and the
   emergent ray. Extend the two rays as shown in the diagram.

6. Use the protractor to draw in the normal at the point where the incident ray enters the prism.

7. Measure the angle between the incident ray and the normal (ie the angle of incidence).
   Make a note of it in your table.

8. Measure the angle of deviation and make a note of it in your table.

9. Repeat the experiment (steps 2 to 8) for a range of 5 angles of incidence between about 30°
   and 70°. Record your results in your table.

10. Plot a graph of angle of deviation against angle of incidence.

11. Write a report describing exactly what you did.

Teacher's Notes *Exercise 7*

Skills tested

Following instructions/manipulation
Presentation of data
Drawing conclusions
Reporting

Previous experience

This is a simple manipulation exercise in which most candidates should be capable of demonstrating positive achievement. More able candidates might be asked questions about the reasons for a straight line graph.

The exercise requires:

that candidates should be familiar with the production of a real image on a screen using a convex lens. They should also be familiar with the method of determining the focal length of a convex lens by focusing the image of a distant object on a screen.

Management strategy

The exercise can be used for individual work in normal practical sessions. Space problems will restrict the number of candidates who can work individually – this will depend on the size and layout of the laboratory.

The exercise cannot be adapted for use as a stations exercise.

Time required

About 20 minutes.

Resources

It is suggested that the lenses are coded with spots of colour paint on the edge, with a different colour for each focal length.

Each candidate will require access to 5 convex lenses with focal lengths between 5 cm and 40 cm.

Longer focal lengths tend to produce an image which is too dim.

Each candidate will also need:

A lens holder
A screen and holder
A metre rule (or half metre)
Graph paper

Marking the assessment

Some or all of the following marking points can be used to develop marking schemes according to the scheme used and the areas assessed.

(a) Assembly and use of apparatus (by observation)
 – suitable position of equipment to obtain image of a distant object
 – image focused
 – care in measurement (use of graph paper on screen or marking screen to enable careful measurement)

(b) Table drawn with
 – columns labelled
 – units included
 – all measurements included
 – all data easily accessible

(c) Graph drawn with
 – correct axes labelled including units
 – suitable scale
 – points correctly plotted
 – best straight line drawn

(d) Statement of relationship between image size and focal length
 – both get bigger
 – directly proportional
 – calculation of constant of proportionality

Exercise 7. Images with Lenses

Before you begin:

Read through all of the instructions carefully.
Make sure that you understand what you have to do.

A convex (converging) lens can produce a real image of a distant object. This image can be focused onto a screen.

You are going to measure the sizes of the images produced by a number of different lenses.

Make each of your measurements as accurately as you can.

Record all of your measurements in a suitable table.

Apparatus   You are provided with:

5 convex lenses (spherical)
A lens holder
A screen with a stand or holder
A metre rule
Graph paper

Check that you have all of the apparatus listed.

You may have to share lenses with other pupils. Your teacher will explain what you should do when you need to change lenses.

Method

1. Set up the screen so that it faces towards a window through which you can see a bright scene.

2. Put the lens holder between the screen and the window. Put a lens on the holder.

3. Carefully adjust the distance between the holder and the screen until you get a sharp, focused image of the scene outside the window.

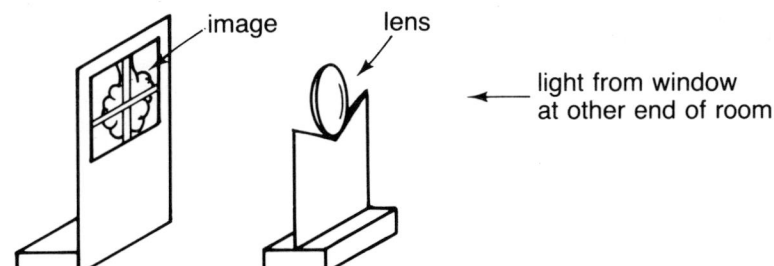

4. Measure accurately the distance between the lens and the screen. Make a note of your measurement in a table.

5. Choose a suitable part of the scene – such as a house or a tree – and measure the size of its image on the screen. Make a note of the measurement in your table. You will measure the same part of the image in the rest of the experiment.

6. If the lens has a label or an identifying mark, make a note of it in the table.

7. Repeat steps 2 to 6 for the other 4 lenses. Record all of your measurements in the table.

8. Plot a graph of the size of the image against the distance from the lens to the screen.

9. What do your results tell you about the connection between the focal length and the size of the image?

10. Write a report describing exactly what you did.

Teacher's Notes

Skills tested

Following instructions/manipulation
Measurement skills – by direct observation
Presentation of data
Drawing conclusions
Reporting

Previous experience

This is a simple cooling curve experiment which is
made slightly more complicated by the need to heat
the liquids in a water bath. The experiment takes a
short time because small quantities of liquid are involved.

The exercise requires:

that candidates should be familiar with heating techniques and with the measurement of volume, time and
temperature. Prior work on heat and heat capacity will
make the physics of the experiment easier to understand, but it is not essential.

Management strategy

Individual work or group work. Working in groups of
two would allow each candidate to perform one half of
the exercise. Results would be pooled for work on the
conclusion.

Time required

About 40 minutes.

Resources

The teacher will need beakers containing water,
labelled A, and liquid paraffin, labelled B.

If paraffin be used, a warning should be given to
keep it away from the Bunsen burner flame or a match
flame. Liquid paraffin is a safer alternative than paraffin
fuel.

Each candidate or group will need:

2 boiling tubes
A beaker
A Bunsen burner, tripod, gauze and bench protection
A stand, boss and clamp
A 0–110 °C thermometer   (contd.)

2 measuring cylinders
A stopwatch
Graph paper

Alternatives and extensions

The apparatus required is so simple that there should
be no problems.

The basic exercise could be developed into a
problem solving/design exercise such as an investigation into the effect of mass on cooling rate.

Marking the assessment

Some or all of the following marking points can be used
to develop marking schemes according to the scheme
used and the areas assessed.

(a) Assembly and use of apparatus (by observation)
    – correct assembly of apparatus
    – use of clean boiling tubes for each liquid
    – safe position and stability of apparatus
    – use of bench mat
    – correct use of Bunsen burner

(b) Measurements (by observation)
    – readings started at correct temperature
    – temperatures correct
    – temperatures taken at correct times

(c) Table drawn with
    – columns labelled
    – units included
    – all measurements included
    – all data easily accessible

(d) Graph drawn with
    – correct axes labelled including units
    – suitable scale
    – points correctly plotted
    – best smooth curves drawn

(e) Recognition of difference in cooling rates between
    two liquids
    Recognition that cooling rates are not constant
    Attempt at comparing cooling rates at a particular
    temperature

Exercise 8. Rates of Cooling

Before you begin:

Read through all of the instructions carefully.
Make sure that you understand what you have to do.

The amount of energy stored by a hot substance depends on
– the substance,
– the mass of the substance,
– the temperature of the substance.

These factors can have an affect on the rate at which the substance cools down.

You are going to study how the use of different substances affects the rate of cooling.

Make all of your measurements in the most accurate way that you can.

Record all of your measurements in a suitable table.

Apparatus   You are provided with:

2 boiling tubes
A beaker
A Bunsen burner, tripod, gauze and bench protection
A stand, boss and clamp
A 0–110 °C thermometer
2 measuring cylinders
A stopwatch
Graph paper

Check that you have all of the apparatus listed.

WEAR GOGGLES DURING THIS EXERCISE.

Method

1. Measure out 50 cm³ of the liquid labelled **A** which your teacher will provide.

2. Pour 50 cm³ of **A** into one boiling tube.

3. Set up the apparatus to heat a beaker of water.

4. Half fill the beaker with water. Stand the boiling tube in the beaker as shown in the diagram.

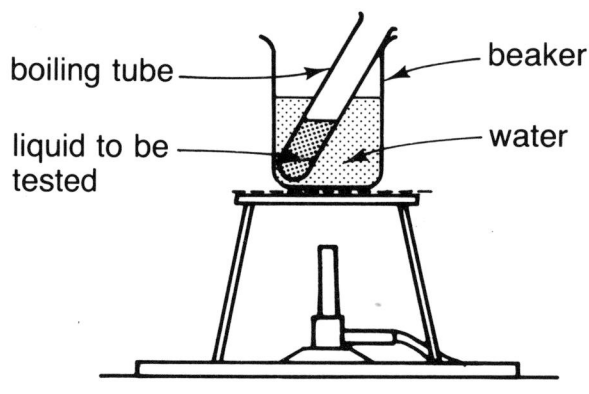

boiling tube — beaker
liquid to be tested — water

5. Carefully heat the beaker of water until the liquid in the boiling tube reaches a temperature of 60 °C.

6. Switch off the Bunsen burner.

7. Use tongs or a strip of wet paper towel to lift the boiling tube from the beaker of water and clamp it vertically with the thermometer in it.

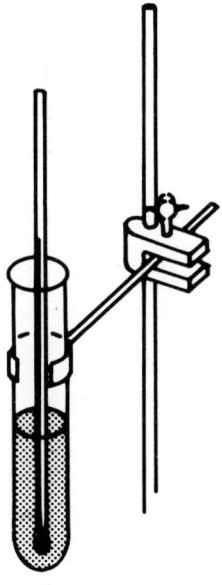

8. Wait until the temperature of the liquid falls to 55 °C.

9. Measure the temperature of the liquid every minute until the temperature reaches 35 °C. Record your results in a table.

10. Pour the liquid **A** down the sink, then rinse and dry the thermometer.

11. Measure 62 cm³ of liquid **B** and pour it into the second boiling tube.

12. Carefully repeat the experiment from steps 2 to 9, but DO NOT pour the liquid away.

13. Plot graphs of temperature against time for each liquid.

14. What do your graphs tell you about the rate of cooling of each liquid?

Teacher's Notes

Skills tested

Following instructions/manipulation
Measurement – by direct observation
Selection of apparatus
Presentation of data

Previous experience

All syllabi contain reference to the thermistor, so here is an exercise which involves measurements using a thermistor.

As the electronics content of syllabi grows, there will be an increasing need for students to measure currents on the milliamp ranges of meters. A digital meter could be used for less able students, but this would exclude the assessment of measuring skills. An alternative exercise which avoids measuring current is suggested below.

The exercise requires:

that candidates should be familiar with connecting an ammeter into a series circuit working from a circuit diagram. They should also be able to measure currents ranging from 0 to 100 mA using the meters available in the centre.

Management strategy

Individual work or in a formal test.

Time required

About 35–40 minutes.

Resources

Each candidate will require a TH-3 thermistor mounted so that it can be immersed in water and connected in a circuit. Such mounted thermistors are available commercially, but a suitable arrangement can be assembled by fixing the thermistor leads to a short strip of wood using drawing pins.

With this arrangement candidates should be provided with a clamp, boss and stand to hold the thermistor mount.

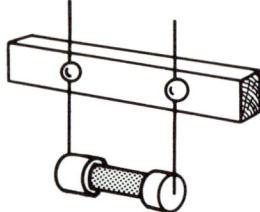

Each candidate will also need:

3 connecting leads, 2 of which should have some means of connecting them to the thermistor – crocodile clips are suitable.

A 6 V d.c. supply (4 × U2 cells)
A 250 cm^3 beaker, water
A Bunsen burner, tripod, gauze and bench protection
A milliammeter with provision for reading 0–10 mA and 0–100 mA
A 0–110 °C thermometer
Graph paper

Alternatives

To avoid the use of the milliammeter a circuit with a resistance of about 1 kΩ in series with the thermistor could be used. A voltmeter would then be used to measure the potential drop across the thermistor. With a 6 V supply, a voltmeter reading 0–1 V and 0–5 V would be suitable.

Question 12 (about resistance) would be inappropriate unless pupils were familiar with the potential divider.

Marking the assessment

Some or all of the following marking points can be used to develop marking schemes according to the scheme used and the areas assessed.

(a) Assembly and use of apparatus (by observation)
  – circuit set up correctly
  – correct assembly of apparatus
  – thermistor immersed
  – safe position and stability of apparatus
  – use of bench mat
  – correct use of Bunsen burner for gentle heating

(b) Measurements (by observation)
  – temperatures correct
  – current readings correct

(c) Table drawn with
  – columns labelled
  – units included
  – all measurements included
  – all data easily accessible

(d) Graph drawn with
  – correct axes labelled including units
  – suitable scale
  – points correctly plotted
  – best smooth curve drawn

(e) Recognition that resistance changes with temperature change.

Exercise 9. The Effect of Temperature on Resistance

Before you begin:

Read through all of the instructions carefully.
Make sure that you understand what you have to do.

A thermistor is a device whose electrical resistance changes with temperature. You are going to measure the current through the thermistor at different temperatures.

The resistance of the thermistor is high, even at high temperatures, so the currents which flow through it are very small. An ammeter which can measure currents in milliamps (mA) will be needed.

Make each of your measurements in the most accurate way that the apparatus will allow.

Record all of your measurements in a suitable table.

Apparatus   You are provided with:

A thermistor
3 connecting leads
A 6 V d.c. supply
A milliammeter which can measure
   0–10 mA and 0–100 mA
A 0–110 °C thermometer
A 250 cm³ beaker
A Bunsen burner, tripod, gauze
   and bench protection
Graph paper

Check that you have all of the
apparatus listed.

Method

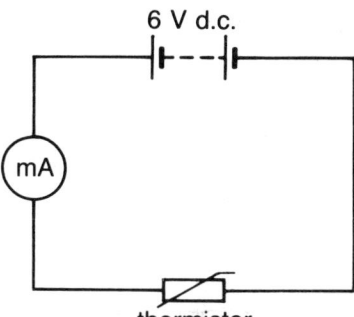

1. Set up the series circuit shown on the right.

2. Measure the current flowing through the thermistor and make a note of it in a table.

3. Measure the temperature of the thermistor and record it in your table.

4. Put enough water in the beaker to cover the thermistor completely. (About ¾ full should be enough.)

5. Set up the beaker on the tripod and gauze and put the thermistor – still connected in the circuit – into the water.

6. Light the Bunsen burner and heat the water gently.

7. When the temperature of the water has risen by about 10 °C, measure the current through the thermistor and the temperature of the water.
   Record them in your table.

8. Continue to measure the current and the temperature at about 10 °C intervals until the water reaches boiling point.
   Record all of your measurements in the table.

9. Measure the current and temperature when the water is boiling and record them in the table.

10. Switch off the Bunsen burner and allow the equipment to cool.

11. Plot a graph of current through the thermistor against temperature using your results.

12. What does your graph tell you about how the resistance of the thermistor varies with temperature?

13. Write a report describing exactly what you did.

Teacher's Notes

Exercise 10

Skills tested

Measurement
- a range of scales should be provided to enable all performance levels to be tested.

Following instructions/manipulation
- the simple nature of the tasks should allow assessment of less able pupils.

Previous experience

The exercises are from a number of different syllabus areas, and candidates should be familiar with the use of the voltmeter before the exercise is carried out.

Management strategy

This is a stations or circus exercise with stepped difficulty. The time required for each exercise is not the same, so it is suggested that stations A and B are combined in one station.

Time required

Allow 8 minutes for (1) and (2) and 8 minutes each for stations (3), (4) and (5).

Resources

Candidate answer sheets should be duplicated. Since candidate records will be used for assessment, the teacher will need to perform the tasks to determine the correct values.

(1)
3 strips of ticker tape of different lengths, marked A, B and C
Metre rule

(2)
A length of wood with a nail in one end
Rubber band
0–10 N newtonmeter
Marks X and Y on the wood so that 3–4 N are needed to stretch the band to X, and 8–9 N are needed to stretch the band to Y.

(3)
Ray lamp with single slit and power supply
Plane mirror and stand
Protractor
Ruler

(4)
6 V d.c. power supply
Rheostat – circuit board type
One $1\,\Omega$ 2.5 W resistor
Two $4.7\,\Omega$ 2.5 W resistors
(The values of the resistors should be obscured by tape)
Ammeter 0–1 A d.c.
Voltmeters 0–1 V, 0–5 V, 0–10 V d.c., marked A, B, C.
Make sure that the p.d. between X and Y is greater than 5 V when the current is 0.5 A.

(5)
5 (or more) identical round objects. They could be ball bearings, pencils, cylinders cut from dowel, lengths of glass rod.
A metre rule
2 set squares

Marking the assessment

Since this exercise assesses measurement skills, each answer should be marked right or wrong according to the values as measured by the teacher.

The exercise is stepped and the level of performance increases with station number for those schemes which use difficulty as a discriminator. If complexity of scale is used, the teacher should choose a suitable range of instruments to ensure that all types of scale specified in the performance descriptions are used.

If the exercise is to be used to assess manipulation/following instructions at the lower levels of performance, the achievement of adequate measurements is a good indicator that the instructions have been followed correctly.

Exercise 10. Measurement Circus – I

In this exercise you will carry out 5 "mini-exercises" which involve measuring.

At each station look at the apparatus provided and read the instructions on this sheet.

When you understand what you must do, make the measurements required as accurately as you can.

---

(1)

On the bench are 3 strips of ticker tape labelled **A**, **B** and **C**. Use the rule to measure the length of each strip as accurately as you can. Record your results on the answer sheet.

---

(2)

On the bench is a piece of wood, an elastic band and a newton meter.

Put the elastic band over the nail and hook the other end onto the newton meter.

Stretch the elastic band until it reaches the line marked **X** and measure the force needed.

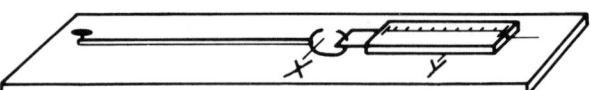

Record your measurement.

Now stretch the elastic band until it reaches the line marked **Y**. Measure the force needed and record your measurement.

---

(3)

On the bench is a ray box and a plane mirror. There is also a ruler and a protractor.
On your answer sheet are 3 lines, AB, CD and IO.

Measure the angle between AB and CD (angle AOC) and write it down.

Stand the mirror with its reflecting surface along AB. Switch on the ray box and shine a ray of light along IO towards the mirror.

Mark the position of the reflected ray carefully, and label it OM.

Now stand the mirror with its reflecting surface along CD and shine the ray at the mirror along IO. Carefully mark the position of the reflected ray and label it ON.

Measure the angle between OM and ON. Is there any connection between the two angles you have measured?

---

(4)

The circuit shown is set up on the bench.

Connect the lead to the battery and adjust the variable resistor until the current through the ammeter is exactly 0.5 A.

Use the voltmeters to measure the potential difference (voltage) between points **X** and **Y**.

Make a note of the voltmeter used in the column on your answer sheet and record the voltage.

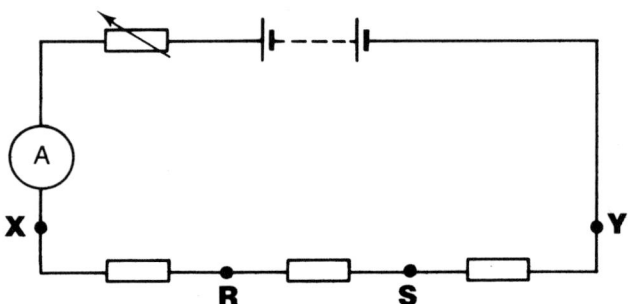

Repeat the procedure to measure the voltages between points X and R, R and S, and S and Y.

---

(5)

On the bench are a number of identical round objects.

Find the diameter of one object as accurately as you can.

Exercise 10. Answer Sheet

Name _____

(1) Length of strip **A** =

Length of strip **B** =

Length of strip **C** =

(2) Force needed to stretch band to **X** =

Force needed to stretch band to **Y** =

(3) Angle AOC =

Angle MON =

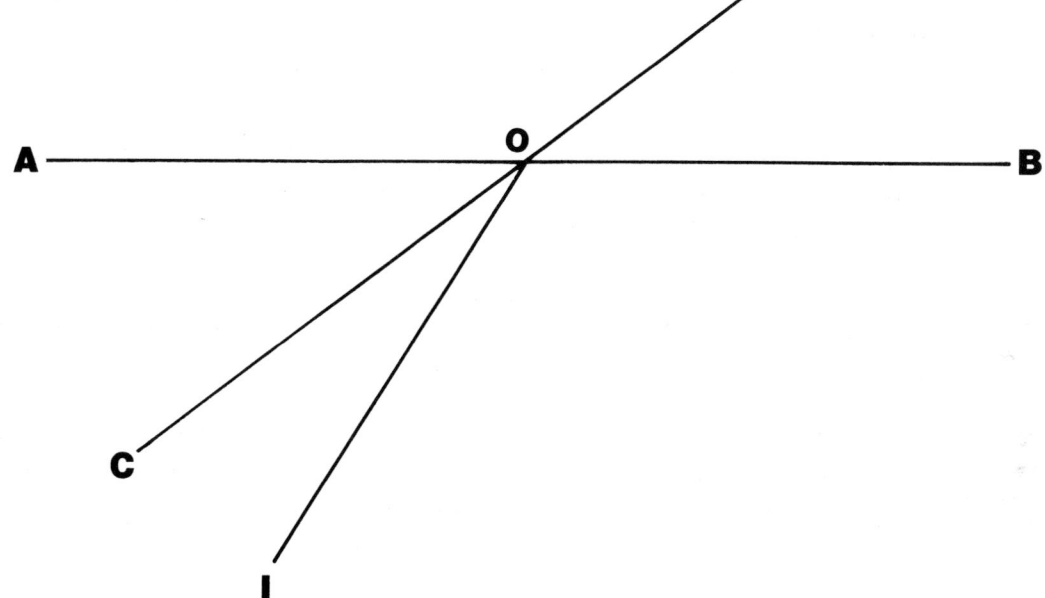

(4)

	Voltage	Meter range used
XY =		
XR =		
RS =		
SY =		

(5) Diameter of one object =

Teacher's Notes

Exercise 11

Skills tested

Measurement
- a range of scales should be provided to enable all performance levels to be tested.

Following instructions/manipulation
- the simple nature of the tasks should allow assessment of less able pupils.

Management strategy

This is another stations exercise with exercises of stepped difficulty. As in Exercise 10, tasks **(1)** and **(2)** should be combined as one station.

Time required

Allow 8 minutes for **(1)** and **(2)**, and 8 minutes each for stations **(3)**, **(4)** and **(5)**. Total 32 minutes.

Resources

Candidate answer sheets should be duplicated. Since candidate records will be used for assessment the teacher will need to perform the tasks to determine the correct values.

(1)

Three 0–110 °C thermometers labelled **A, B, C,** clamped vertically. **B** and **C** should have a piece of filter paper wrapped around the bulb, held with an elastic band. The filter paper dips into 100 cm³ beakers containing for **B**, water, and for **C**, colourless alcohol to produce "wet-thermometers".

Check the readings on each thermometer during the session.

(2)

A beaker with a strip of paper stuck to the outside with two horizontal marks labelled **A** and **B**. The position of the marks should be such that about 70 cm³ are required to fill the beaker to **A** and about 180 cm³ are needed to fill it to **B**.
Another beaker containing at least 200 cm³ of water.
A range of measuring cylinders suitable for measuring the water volumes. Each cylinder should be labelled with a different letter.

(3)

This station should be set up next to a sink.
A plastic tray about 7 cm deep and 50 cm long: the type used for storage is suitable.
Beaker for adding water to the tray
Stopwatch
Metre rule
Cloth for spillage

(4)

6 V d.c. power supply
Mounted 5 KΩ potentiometer labelled **C**
1 kΩ resistor
2.2 kΩ resistor
(Resistance substitution boxes could be used with the controls taped)
Two 0–5 V d.c. voltmeters labelled V_1 and V_2
Connecting leads

Check the readings of V_1 and V_2 to determine the ratio V_1/V_2.

(5)

A wooden burette stand
2 bar magnets – one labelled with a number 2
Stopwatch
Metre rule

Hang one of the bar magnets from the stand by means of a length of thread so that the magnet is horizontal and can oscillate as a torsion pendulum. A stirrup or loop of thread taped to the magnet is a suitable method of suspending the magnet.

The magnet should be about 15cm above the bench. Adjust the height so that the second magnet decreases the period by a factor of about 2 – the height will depend on the magnets used.

Marking the assessment

Since this exercise assesses measurement skills, each answer should be marked right or wrong according to the values as measured by the teacher.

The exercise is stepped and the level of performance increases with station number for those schemes which use difficulty as a discriminator. If complexity of scale is used, the teacher should choose a suitable range of instruments to ensure that all types of scale specified in the performance descriptions are used.

Exercise 11. Measurement Circus – II

In this exercise you will carry out 5 "mini-exercises" which involve measuring.

At each station look at the apparatus provided and read the instructions on this sheet.

When you understand what you must do, make the measurements required as accurately as you can.

---

(1)

Three thermometers, **A**, **B** and **C**, are set up on the bench.

Read the temperature on each of the thermometers and record your measurements on the sheet.

---

(2)

On the bench is a range of measuring cylinders, each labelled with a letter.

Pour water into the beaker until it reaches level **A** exactly.

Measure the volume of this water and record your answer.

Now add water until it reaches level **B** exactly. Measure the volume of this water and record your answer.

In each case, select the best measuring cylinder to make the measurement.

---

(3)

Pour water into the plastic tray to a depth of exactly 1 cm.

Send a wave down the tray by lifting one end slightly, then letting it back down to the bench.

Once you have succeeded, measure the time taken for the wave to travel from one end of the tray to the other and record your measurement.

Repeat the experiment for water depths of exactly 2 cm and 4 cm.

Carefully empty the water into the sink. Measure the length of the inside of the tray as accurately as you can.

---

(4)

The circuit shown is set out on the bench. Complete the circuit by connecting the flying lead to the +6 V terminal on the battery.

Adjust the knob on the controller **C** until the reading on V_1 is exactly 5 V. Measure the reading on voltmeter V_2 and record it in the table.

Set the reading on V_1 to four more different values, and for each one, record the reading on V_1 and V_2 in the table.

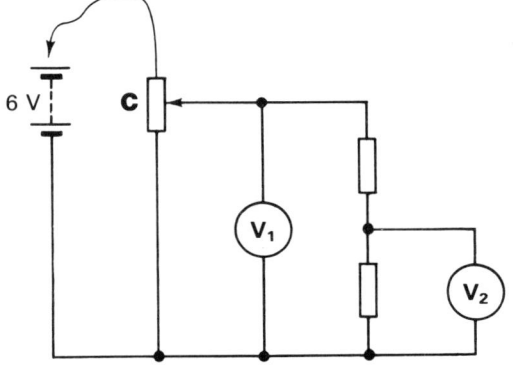

Disconnect the flying lead from the battery when you have finished.

---

(5)

On the bench is a bar magnet hung so that it can twist freely, and a second magnet labelled 2. Place magnet **2** as far away as you can from the first magnet.

Set the hanging magnet into oscillation by twisting it and then letting it go.

Measure the time for one complete oscillation as accurately as you can.

Now put the second magnet on the bench so that it is directly underneath the hanging magnet. Use the rule to help you do this.

Set the hanging magnet into oscillation again and measure the time for one complete oscillation as accurately as you can.

Exercise 11. Answer Sheet

Name _____

(1) Temperature of thermometer **A** =

Temperature of thermometer **B** =

Temperature of thermometer **C** =

(2) Volume of water to level **A** =

Measuring cylinder used:

Volume of water to level **B** =

Measuring cylinder used:

(3)

Depth of water	Time
1 cm	
2 cm	
4 cm	

Length of inside of tray =

(4)

V_1	V_2
5 V	

(5)

Time for 1 oscillation without 2nd magnet	Time for 1 oscillation with 2nd magnet

Teacher's Notes

Exercise 12(a)

Skills tested

All skill areas.

Previous experience

This exercise has been included to show how a traditional experiment can be adapted to turn it into a problem-solving experiment.

The exercise requires:
that candidates should be familiar with techniques of volume and mass measurement for determination of density.

Management strategy

Individual work in a formal test situation.

Time required

About 40 minutes – time will depend on the availability of balances.

Resources

The candidate will need a set of 5 numbered samples of different materials with a suitable range of densities.

A table of densities including the densities of the 5 samples plus some others should be provided.

The items in the table should be identifiable only by letter – as in the table in Exercise 12(b).

Candidates who are unable to proceed could be given a copy of Exercise 1 to help them.

Marking the assessment

Some or all of the following marking points can be used to develop marking schemes according to the scheme used.

(a) Identification of the best method (density determination).

(b) Choice of appropriate measuring instruments to enable sufficiently accurate measurements to be made.

(c) Care in making measurements and accuracy of results.

(d) Identification of samples.

Teacher's Notes

Exercise 12(b)

Skills tested

This is a more difficult exercise than Exercise 12(a) because candidates must decide on the most appropriate physical property for identification.

The particular skills involved will depend upon the candidate's strategy for solving the problem and carrying out the experiment.

Resources

The substances in the table are:

A Aluminium
B Copper
C Gold
D Iron
E Mercury
F Zinc
G Glass
H Sulphur
I Paraffin Wax
J Turpentine

Five samples of substances from the table should be provided, each one identified by a letter. It is suggested that mercury is not used for the experiment even though its properties are included in the table.

Marking the assessment

Some or all of the following marking points can be used to develop marking schemes according to the scheme used.

(a) Identification of suitable methods (these will depend on the samples provided).

(b) Choice of appropriate measuring instruments to enable sufficiently accurate measurements to be made.

(c) Care in making measurements and accuracy of results.

(d) Identification of samples.

Exercise 12(a). Using Density to Identify Substances

One practical use of the connection between the mass of a piece of a substance and its volume is in helping to identify unknown rocks and minerals. Geologists measure a quantity called DENSITY (the mass of 1 cm^3 of a substance) and compare the result with a table of the densities of known minerals.

Design an experiment which will enable you to identify each of the samples provided.

Write down details of the experiment which you intend to carry out. Make sure you say:

> ★ which apparatus you will use
>
> ★ what you will do
>
> ★ how you will identify each sample.

DO NOT CONTINUE UNTIL YOUR PLAN HAS BEEN CHECKED BY YOUR TEACHER.

If you are unable to plan a suitable experiment or do not know where to begin, then see your teacher. You will be given a sheet which should help you to begin.

When your teacher tells you, you can carry out your experiment. If you decide to change how you carry out part of your experiment you may do so, but make sure that you explain the reason for your decision when you write your report.

Write a report of your experiment. Say what you did and how you did it. Include a table of your measurements and use your results to identify each of the samples.

Exercise 12(b). Identifying Substances

Here is a table of physical properties of some substances.

Substance	Density (g/cm^3)	Melting point (°C)	Specific heat capacity (J/kg K)	Electrical conductivity
A	2.7	660	900	good
B	8.9	1083	380	good
C	19.3	1063	130	good
D	7.9	1539	460	good
E	13.6	−38.7	140	good
F	7.1	419	380	good
G	2.6	1127	670	poor
H	2.0	113	730	poor
I	0.9	57	2900	poor
J	0.9	−10	1760	poor

You are provided with samples of five different substances.

Design a set of simple experiments using normal laboratory apparatus which will enable you to identify each sample with one of the substances in the table.

Write down details of the experiments which you intent to carry out. Make sure you say:

★ which apparatus you will use

★ what you will do

★ how you will identify each sample from the results of your tests.

DO NOT CONTINUE UNTIL YOUR PLAN HAS BEEN CHECKED BY YOUR TEACHER.

When your teacher tells you, you can carry out your experiments.

Write a report of your experiments. Say what you did and how you did it. If you change any of your plan, explain the reason for your change in your report.

Include tables of your measurements and conclusions in your report.

Teacher's Notes   *Exercise 13*

Management strategy

This is a simple problem which involves a task which should be appropriate to most pupils. The major problem is the control of variables. The statement of the problem is left blank so that teachers can add the variable in the space provided. The diameter of the cylinder or the height of the cylinder are suggested.

Pupils who have difficulty designing a suitable experiment can be provided with a range of paper tubes of different heights and diameters. They will still need to select which are the most suitable for their particular investigation. They can also be supplied with a range of apparatus from which to choose suitable items.

Time required

Pupils should be able to complete the practical part of the exercise in about 30 minutes.

Resources

Pupils will need sheets of thin paper such as bank copy paper and sticky tape for making the tubes.

In addition, supplies of masses, metre rules, scissors and squares of hardboard about 8 cm square will probably be required.

For pupils with difficulty designing the experiment:

Selection of paper tubes.
Squares of thick carboard, hardboard and steel or other heavy platform materials. Some squares should have a hole of about 1 cm diameter in the centre.

A piece of wood about 10 cm square and 1 cm thick with a hook in the centre.
Newton-meters 0–10 N and 0–50 N
Stand, clamp and boss
Range of slotted masses to provide a load up to 10 N
String

Marking the assessment

Some or all of the following marking points can be used to develop marking schemes according to the scheme used.

(a) Identification of the variables:
   – diameter of tube
   – length of tube
   – load

(b) Control of variables (these will depend on the problem set).

(c) Range of variables used.

(d) Choice of appropriate method of loading tubes.

(e) Care in making measurements and accuracy of results.

(f) Presentation and handling of results.

(g) Conclusions reached.

(h) Suggestions for improvements to experiment.

Exercise 13(a). Crushing Tubes

Steel tubes (which can be used in structures such as cranes and oil rigs) sometimes collapse by crushing. Various factors can affect the load which a tube can support. You are going to design and carry out an experiment to study the effect of one of them.

You can use tubes made from paper instead of steel.

You are provided with sheets of thin paper and sticky tape which you can use to make the tubes.

> DESIGN AN EXPERIMENT TO FIND OUT HOW THE FORCE NEEDED TO CRUSH A PAPER TUBE DEPENDS ON
>
> --

Write out details of the experiment which you intend to carry out. Make sure you say:

- ★ what apparatus you will use

- ★ what measurements you will make

- ★ what you will do to make sure your experiment is a fair test.

DO NOT CONTINUE UNTIL YOUR PLAN HAS BEEN CHECKED BY YOUR TEACHER.

If you are unable to plan a suitable experiment or do not know where to begin, see your teacher. You will be given information which should help you to begin.

You can carry out your experiment when your teacher tells you. If you decide to change your experiment you may do so, but be sure to explain the reason for your decision when you write your report.

Write a report of your experiment. Say what you did and how you did it. Include a table of your results and conclusions in your report.

Exercise 13(b)

If you are unable to design a suitable experiment the following notes may help you.

Your teacher will provide you with a selection of paper tubes and some apparatus. The apparatus can be used to measure the tubes and to measure the size of the force on the tubes.

1. Remember to use only the tubes which will make your experiment a fair test.

2. There are a number of ways of applying a load to the tube. The simplest method is usually the best.

3. You should apply the load steadily and increase it gently. Sudden increases in load might affect your results.

Teacher's Notes

Management strategy

This is a very simple problem. Pupils can investigate how the height of the pile depends on:

(a) the size of the particles

(b) the volume of material

or how the diameter of the pile depends on:

(a) the size of the particles

(b) the volume of material.

Spaces are left in the pupil sheet to allow the teacher a choice.

If the pupil has problems, Exercise 14(b) provides some clues.

Time required

The practical part of the exercise should be completed in 30 minutes.

Resources

Pupils will require a set of samples of "runny solids" with different particle sizes. Suggestions are dry salt, dry sieved sand, sawdust, granulated sugar, lentils, rice, split peas.

It is likely that they will also require measuring cylinders, beakers, metre or half metre rules, graph paper.

Marking the assessment

Some or all of the following marking points can be used to develop marking schemes according to the scheme used.

(a) Identification of the variables
   – diameter of pile
   – height of pile
   – size of particles
   – volume of material
   – height of pouring

(b) Control of variables (these will depend on the problem set).

(c) Ranges of variables used.

(d) Care in making measurements (including repeating readings) and accuracy of results.

(e) Presentation and handling of results.

(f) Conclusions reached.

(g) Suggestions for improvements to experiment.

Exercise 14(a). Pouring Powders

There are many everyday situations in which large amounts of loose substances such as rock, coal and grain are loosely piled up. Piles of loose material tend to settle in stable heaps. If they are too steep they can collapse and bury people, so the behaviour of such heaps is important.

You can use powders or small grains of material to model the behaviour of heaps.

You are provided with samples of different substances.

DO NOT MIX THEM UP.

> DESIGN AN EXPERIMENT TO FIND OUT HOW THE ....
>
> OF A HEAP DEPENDS ON
>
> ....

Write out details of the experiment which you intent to carry out. Make sure you say:

- ★ what apparatus you will use
- ★ what measurements you will make
- ★ what you will do to make sure your experiment is a fair test

DO NOT CONTINUE UNTIL YOUR PLAN HAS BEEN CHECKED BY YOUR TEACHER.

If you are unable to plan a suitable experiment or do not know where to begin, see your teacher. You will be given information which should help you to begin.

You can carry out your experiment when your teacher tells you. If you decide to change your experiment you may do so, but be sure to explain the reason for your decision when you write your report.

Write a report of your experiment. Say what you did and how you did it. Include a table of your results and conclusions in your report.

Exercise 14(b)

If you are unable to design a suitable experiment, the following notes may help you.

1. You can use a measuring cylinder to measure the volume of material.

2. A cone-shaped heap can be produced by pouring the material in a steady stream from a beaker. Hold the beaker so that the lip is just above the top of the heap as it builds up.

3. Pouring the pile onto graph paper will help you to measure the diameter of the heap if you need to do this.

4. Pour the heap on to a sheet of paper so that it can easily be returned to its beaker.

Teacher's Notes

Management strategy

This is another very simple problem which should be capable of a resonable solution by pupils of all levels of ability. The range of skills required is limited and quite simple so it is appropriate for less able pupils.

Pupils who find problems with designing an experiment can be provided with Exercise 15(b).

The simple conclusion will be a statement of the time taken for a given quantity of water to filter at the beginning and the end of the process. More able pupils will be able to calculate flow rates or discuss the results in terms of the gradient of a graph of volume against time.

Time required

The practical part of the exercise can be carried out in 30 minutes.

Resources

Pupils will need samples of a dry insoluble powder. Dried coffee grounds, sawdust or sand would be suitable.

It is likely that they will require measuring cylinders, stopwatches, filter funnels and filter papers, stands and clamps, and a measure such as a spoon for measuring the powder.

Marking the assessment

Some or all of the following marking points can be used to develop marking schemes according to the scheme used.

(a) Identification of the variables:
   – volume of liquid
   – volume of powder
   – time
   – state of powder (wet or dry)
   – state of filter paper (wet or dry)

(b) Control of variables:
   – volume of liquid used
   – quantity and state of powder
   – state of filter paper

(c) Range of variables used.

(d) Choice of appropriate method of measuring flow rate.

(e) Care in making measurements (including repeating readings) and accuracy of results.

(f) Presentation and handling of results.

(g) Conclusions reached.

(h) Suggestions for improvements to experiment.

Exercise 15(a). Filtering

You may have noticed when filtering liquids in chemistry that the liquid seems to run through more slowly at the end of the process than at the beginning.

I have noticed the same problem when making coffee by pouring boiling water through a filter containing finely ground coffee beans.

You are provided with a dry sample of an insoluble powder.

> DESIGN AN EXPERIMENT WHICH WILL ENABLE YOU TO MAKE AN ACCURATE TEST OF HOW THE RATE OF FILTERING VARIES WITH THE AMOUNT OF WATER WHICH HAS RUN THROUGH THE FILTER

Write out details of the experiment which you intend to carry out. Make sure you say

★ what apparatus you will use

★ what measurements you will make

★ what you will do to make sure your experiment is a fair test.

DO NOT CONTINUE UNTIL YOUR PLAN HAS BEEN CHECKED BY YOUR TEACHER.

If you are unable to plan a suitable experiment or do not know where to begin, see your teacher. You will be given information which should help you to begin.

You can carry out your experiment when your teacher tells you. If you decide to change your experiment you may do so, but be sure to explain the reason for your decision when you write your report.

Write a report of your experiment. Say what you did and how you did it. Include a table of your results and conclusions in your report.

Exercise 15(b)

If you have difficulty in designing a suitable experiment the following notes may help you.

1. Use a filter funnel and filter paper.

2. Measure the powder into the filter with the measure provided.

3. Pour water into the funnel and allow it to run into a measuring cylinder.

4. Use a stopwatch to time the water running into the cylinder.

5. If you wish to repeat the experiment use a dry filter paper and a fresh dry sample of powder.

Teacher's Notes *Exercise 16*

Management strategy

This exercise offers a range of investigations. Pupils can be provided with different containers, such as beakers and pans or containers with and without lids; or they can be presented with the problem of investigating the effect of mass on rate of temperature rise. In the first two instances the pupils should be provided with the containers before they start to design the experiment.

The spaces on the pupil sheet allow the teacher to alter the instructions to suit the problem set.

Time required

This will depend on the size of the containers provided and the efficiency of the Bunsen burners. However, it should be possible to complete the exercise in about 40 minutes.

Resources

The teacher will need to collect a range of suitable containers and lids. It is likely that the rest of the equipment requested will be standard laboratory apparatus – measuring cylinders, thermometers, beakers, stopwatches, Bunsen burners, tripod, gauze and bench protection.
Graph paper should be available.

Marking the assessment

Some or all of the following marking points can be used to develop marking schemes according to the scheme used.

(a) Identification of the variables:
   – mass of water
   – container used
   – time
   – temperature
   – setting of heater

(b) Control of variables (these will depend on the problem set).

(c) Range of variables used.

(d) Care in making measurements (including repeating readings) and accuracy of results.

(e) Presentation and handling of results.

(f) Conclusions reached.

(g) Suggestions for improvements to experiment.

Exercise 16(a). The Best Way to Boil Water

You may have noticed that the time taken to bring a pan of water to boiling on a cooker can depend on a number of things including – the amount of water, the size of pan used and whether or not you have a lid on the pan.

You are provided with ....

> **DESIGN AN EXPERIMENT TO FIND OUT HOW THE TIME TAKEN TO BOIL WATER USING A BUNSEN BURNER DEPENDS UPON**
>
> ....

Write out details of the experiment which you intend to carry out. Make sure you say:

★ what apparatus you will use

★ what measurements you will make

★ what you will do to make sure your experiment is a fair test

DO NOT CONTINUE UNTIL YOUR PLAN HAS BEEN CHECKED BY YOUR TEACHER.

If you are unable to plan a suitable experiment or do not know where to begin, see your teacher. You will be given information which should help you to begin.

You can carry out your experiment when your teacher tells you. If you decide to change your experiment you may do so, but be sure to explain the reason for your decision when you write your report.

Write a report of your experiment. Say what you did and how you did it. Include a table of your results and conclusions in your report.

Exercise 16(b)

If you are unable to design a suitable experiment the following notes may help you.

1. You must vary only the factor under investigation during your experiment. All of the other factors – including the setting of the gas tap and the air hole of the Bunsen burner – should NOT be changed during the experiment.

2. You will probably need to measure the volume or mass of water which you heat.

3. It is more accurate to measure the temperature of the water at intervals during the heating rather than just the time taken for it to boil.

Teacher's Notes

Management strategy

This exercise is similar in nature to Exercise 16 and could be used as an alternative exercise in the same session.

Time required

The practical part of the exercise should be completed in about 30 minutes.

Resources

The teacher will need to amass a suitable stock of candles (see below), night lights and other energy sources.

Pupils are likely to require beakers, test-tubes, measuring cylinders, thermometers and stopwatches.

Suitable sources for the experiment are candles of different shapes and sizes, night lights, spirit burners and solid fuel used for camping stoves. Alternatively, home-made "candles", using oil or fat in an evaporating basin with a string wick, could be used.

Alternative exercise

Able pupils could be given the task of comparing the energy content of different fuels.

Marking the assessment

Some or all of the following marking points can be used to develop marking schemes according to the scheme used.

(a) Identification of a suitable method of estimating rate of energy production.

(b) Identification of the variables (if the method involves rate of temperature rise produced in a sample of water):
 – temperature
 – time
 – mass of water
 – container used
 – type of fuel
 – method of burning

(c) Control of variables:
 – mass of water used
 – container used

(d) Range of variables used.

(e) Care in making measurements (including repeating readings) and accuracy of results.

(f) Presentation and handling of results.

(g) Conclusions reached.

(h) Suggestions for improvements to experiment.

Exercise 17(a). Testing Fuels

Different fuels and different ways of burning fuels produce energy at different rates.

You are provided with two samples of fuel or ways of burning fuel.

> DESIGN AN EXPERIMENT WHICH WILL ENABLE YOU TO DECIDE WHICH OF YOUR SAMPLES PRODUCES HEAT ENERGY AT THE HIGHEST RATE.

Write out details of the experiment which you intend to carry out. Make sure you say:

- ★ what apparatus you will use

- ★ what measurements you will make

- ★ what you will do to make sure your experiment is a fair test

DO NOT CONTINUE UNTIL YOUR PLAN HAS BEEN CHECKED BY YOUR TEACHER.

If you are unable to plan a suitable experiment or do not know where to begin, see your teacher. You will be given information which should help you to begin.

You can carry out your experiment when your teacher tells you. If you decide to change your experiment you may do so, but be sure to explain the reason for your decision when you write your report.

Write a report of your experiment. Say what you did and how you did it. Include a table of your results and conclusions in your report.

Exercise 17(b)

If you have difficulty in designing a suitable experiment the following notes may help you.

1. You can compare the energy output of the fuels by using them to heat water and comparing the rates of temperature rise.

2. You will need to use a suitable quantity of water for your tests. Too much water will take a long time to heat up. If you use a tiny amount of water it will heat up too quickly.

3. Measure the temperature of the water at regular intervals of time so that you can get an accurate measure of the rates of temperature rise.

Teacher's Notes *Exercise 18*

Management strategy

This exercise can be used to test the effect of

(a) distance
(b) the mass of steel
(c) the number of sheets of paper between the magnet and a piece of steel

on the force of attraction.

It is so straightforward that it is unlikely that candidates will require any additional help.

Pupils should be provided with a piece of steel and a bar magnet at the beginning of the exercise. Taping a loop of string to the magnet might provide a clue for less able candidates.

Time required

This exercise could be completed in about 20 minutes.

Resources

Candidates will require a bar magnet and a piece of steel, such as the base of a clamp stand.

In addition they may well request:
newton meters
balances
paper and scissors
rules
masses
different-sized pieces of steel

Marking the assessment

Some or all of the following marking points can be used to develop marking schemes according to the scheme used.

(a) Identification of the variables (these will depend on the problem set and the range of equipment and magnets provided).

(b) Control of variables (these will depend on the problem set).

(c) Range of variables used.

(d) Choice of appropriate method of measuring force of attraction.

(e) Care in making measurements (including repeating readings) and accuracy of results.

(f) Presentation and handling of results.

(g) Conclusions reached.

(h) Suggestions for improvements to experiment.

Exercise 18. Measuring Magnetic Forces

The force of attraction between a bar magnet and a piece of steel can depend on a number of factors.

It depends on the distance between the magnet and the steel.
It is also affected by placing sheet of card or paper between the steel and the magnet.
It may be affected by the size of the piece of steel.

DESIGN AN EXPERIMENT TO DETERMINE HOW THE FORCE OF ATTRACTION BETWEEN A BAR MAGNET AND A PIECE OF STEEL DEPENDS ON

Write out details of the experiment which you intend to carry out. Make sure you say:

★ what apparatus you will use

★ what measurements you will make

★ what you will do to make sure your experiment is a fair test

DO NOT CONTINUE UNTIL YOUR PLAN HAS BEEN CHECKED BY YOUR TEACHER.

If you are unable to plan a suitable experiment or do not know where to begin, see your teacher. You will be given information which should help you to begin.

You can carry out your experiment when your teacher tells you. If you decide to change your experiment you may do so, but be sure to explain the reason for your decision when you write your report.

Write a report of your experiment. Say what you did and how you did it. Include a table of your results and conclusions in your report.

Teacher's Notes *Exercise 19*

Management strategy

This is a relatively simple exercise although the problems of measuring the small currents involved may restrict its use to the more able.

The more observant students will notice the problem of keeping the illumination level steady. Movements near to the apparatus will cause fluctuations in the resistance of the LDR and the most able pupils will probably take steps to avoid the problem.

There are a number of factors which might be investigated. Possibilities include the effect of colour of filter, the effect of number of filters or, if pupils are provided with a range of solutions of copper sulphate or potassium manganate (VII), the effect of concentration on resistance.

Time required

This will depend on the factor to be studied. Using filters will produce a shorter exercise than using solutions.

Resources

Pupils should be provided with an ORP 12 LDR mounted so that it can be connected in a circuit.

Meters, leads and 1.5 V cells in cell holders should be available for pupils to make rough trials.

Supplies of coloured filters, solutions of different concentrations (labelled) and light sources should also be available.

Pupils who are unable to begin could be provided with the LDR connected in a circuit with a milliammeter so that they can begin trial investigations on the effect of illumination on the current.

Marking the assessment

Some or all of the following marking points can be used to develop marking schemes according to the scheme used.

(a) Identification of the variables (these will depend on the problem set).

(b) Control of variables (these will depend on the problem set).

(c) Choice of appropriate method of measuring resistance.

(d) Range of variables used.

(e) Care in making measurements (including repeating readings) and accuracy of results.

(f) Presentation and handling of results.

(g) Conclusions reached.

(h) Suggestions for improvements to experiment.

Exercise 19(a). Investigating a Light Dependent Resistor

A light dependent resistor (LDR) is a component whose electrical resistance depends on the **brightness** of the light which falls on it. The change in its resistance also depends on the **colour** of the light which falls on it.

The resistance of an LDR can be used to measure the brightness of light, or how much light is absorbed, by passing through a filter or a number of filters.

> DESIGN AN EXPERIMENT TO FIND OUT HOW THE RESISTANCE OF AN LDR DEPENDS ON
>
> ....

Note The resistance of the LDR is quite high, even in daylight, so you will need to measure quite small currents.

Write out details of the experiment which you intend to carry out. Make sure you say:

★ what apparatus you will use

★ what measurements you will make

★ what you will do to make sure your experiment is a fair test

DO NOT CONTINUE UNTIL YOUR PLAN HAS BEEN CHECKED BY YOUR TEACHER.

If you are unable to plan a suitable experiment or do not know where to begin, see your teacher. You will be given information which should help you to begin.

You can carry out your experiment when your teacher tells you. If you decide to change your experiment you may do so, but be sure to explain the reason for your decision when you write your report.

Write a report of your experiment. Say what you did and how you did it. Include a table of your results and conclusions in your report.

Exercise 19(b)

If you are unable to design a suitable experiment, your teacher will provide you with the equipment to make the circuit shown.

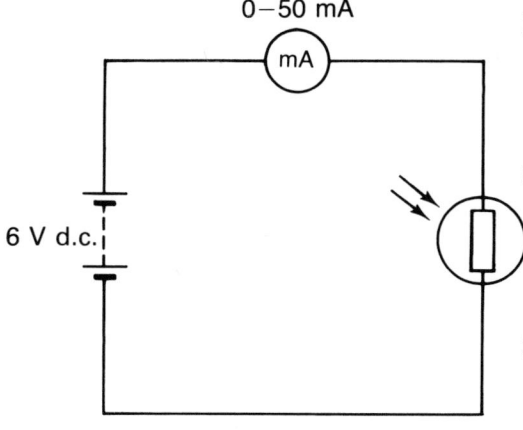

0–50 mA

6 V d.c.

Method

1. Set up the circuit and watch what happens to the current as you put your hand over the LDR.

2. For your experiment you will need a steady source of light. You will need to make sure that changes in the light in the laboratory do not affect your apparatus.

3. With the circuit left, the current is a good measure of the resistance of the LDR.

Previous experience

Although this is a simple investigation to perform, it requires an understanding of resistance and potential for the planning. It is probably more suitable for the more able students.

Management strategy

Providing students with a suitable circuit, so that they can investigate the effect of the length of the resistance wire on current or p.d., should furnish them with clues as to how to solve the problem.

With careful measurements it is possible to determine the position of the junction to within one millimetre.

Time required

About 30 minutes.

Resources

The "problem" consists of a 100 cm length of 0.28 mm (32 s.w.g.) nickel–chromium wire fastened along a metre rule so that electrical connections can be made to its ends.

A piece of insulated copper wire is soldered to the resistance wire to produce a T-junction and the position of the junction is obscured. A length of rubber tube taped to the rule so that it cannot be moved is suitable. The end of the wire at the zero of the rule is labelled **A**.

Pupils will also require 6 V batteries or a suitable d.c. power supply and access to connecting leads and meters.

A suitable "jockey", such as a small screwdriver with a lead, should be provided to act as a flying lead for touching the resistance wire.

Marking the assessment

Some or all of the following marking points can be used to develop marking schemes according to the scheme used.

(a) Identification of a suitable method of solution.

(b) Identification of variables:
  – length of wire
  – potential difference
  – setting of power supply

(c) Control of variables:
  – setting of power supply

(d) Range of variables used.

(e) Care in making measurements (including repeating readings) and accuracy of results.

(f) Presentation and handling of results.

(g) Conclusions reached.

(h) Suggestions for improvements to experiment.

Exercise 20(a). Electrical Junction Puzzle

The early development of ways of measuring electrical resistance was spurred on by attempts to find faults in telegraph wires without inspecting the wire along its entire length.

By measuring the resistance of a pair of wires very accurately, Sir Charles Wheatstone was able to find the position of a short circuit between them.

In this experiment you are provided with a long piece of resistance wire which has a piece of copper wire joined to it somewhere along its length. The junction is hidden, so you cannot find out where it is directly.

> DESIGN AN EXPERIMENT WHICH WILL ENABLE YOU TO DETERMINE THE EXACT DISTANCE BETWEEN THE END OF THE RESISTANCE WIRE **A** AND THE JUNCTION.

Write out details of the experiment which you intend to carry out. Make sure you say:

- ★ what apparatus you will use

- ★ what measurements you will make

- ★ what you will do with your measurements to answer the problem

DO NOT CONTINUE UNTIL YOUR PLAN HAS BEEN CHECKED BY YOUR TEACHER.

If you are unable to plan a suitable experiment or do not know where to begin, see your teacher. You will be given information which should help you to begin.

You can carry out your experiment when your teacher tells you. If you decide to change your experiment you may do so, but be sure to explain the reason for your decision when you write your report.

Write a report of your experiment. Say what you did and how you did it. Include a table of your results and conclusions in your report.

Exercise 20(b)

If you are unable to design a suitable experiment, the following notes may help you.

Method

1. Set up the apparatus as shown in the picture.

2. Touch the resistance wire with the screwdriver and notice what happens to the pointer of the voltmeter.

low voltage power supply

screwdriver

resistance wire

copper wire

voltmeter

A

3. Move the point of contact along the resistance wire and notice what happens to the reading on the voltmeter.

4. Touch the end of the copper wire and notice what happens to the pointer of the voltmeter.

References

The following publications are referred to in the text or are recommended for further reading.

ASSESSMENT OF PERFORMANCE UNIT (1984), *Science Assessment Framework Age 13 and 15*, DES, London (APU Science Report for Teachers: 2)

ASSESSMENT OF PERFORMANCE UNIT (1985), *Science at Age 15*, DES, London (APU Science Report for Teachers: 5)

ASSESSMENT OF PERFORMANCE UNIT (1985), *Practical Testing at Ages 11, 13 and 15*, DES, London (APU Science Report for Teachers: 6)

ASSESSMENT OF PERFORMANCE UNIT (1985), *Science in Schools: Ages 13 and 15*, DES, London (APU Research Report: 3)

BULMAN L. (1985), *Teaching Language and Study Skills in Secondary Science*, Heinemann, London

EGLEN J.R. and KEMPA R.F. (1974), "Assessing manipulative skills in practical chemistry", *School Science Review*, **56,** 261–63

KEMPA R.F. (1986), *Assessment in Science*, Cambridge University Press, Cambridge

SECONDARY EXAMINATIONS COUNCIL (1986), *Science GCSE: A Guide for Teachers*, Open University Press, Milton Keynes

SOLOMON J. (1980), *Teaching Children in the Laboratory*, Croom Helm, London

WOOLNOUGH B. and ALLSOP T. (1985), *Practical Work in Science*, Cambridge University Press, Cambridge